T H E
DEVIL'S
DELUSION

THE

DEVIL'S

DELUSION

Atheism and Its Scientific Pretensions

DAVID BERLINSKI

BASIC
BOOKS

A Member of the Perseus Books Group
New York

Copyright © 2009 by David Berlinski

Published by Basic Books,
A Member of the Perseus Books Group

Published in the United States in 2008 by Crown Forum, an imprint of the
Crown Publishing Group, a division of Random House, Inc.

Books published by Basic Books are available at special discounts for bulk
purchases in the United States by corporations, institutions, and other
organizations. For more information, please contact the Special Markets
Department at the Perseus Books Group, 2300 Chestnut Street, Suite 200,
Philadelphia, PA 19103, or call (800) 810-4145, ext. 5000, or e-mail
special.markets@perseusbooks.com.

A CIP catalog record for this book is available from the Library of
Congress.

Library of Congress Control Number: 2009931847

ISBN: 978-0-465-01937-3

LSC-C

20 19 18 17 16 15

To the memory of
my maternal grandfather

S AMUEL G OLDFEIN

15.1.1877 Pruzani
auf den Transportlisten von 19.9.42 um 17.2.43 gestrichen
am 27.2.43 nach Dresden
am 29.3.43 nach Theresienstadt
am 18.12.43 nach Auschwitz deportiert
in Auschwitz verschollen

CONTENTS

He must have a long spoon that must eat with the devil.

❧

—SHAKESPEARE, *The Comedy of Errors*

PREFACE TO THE SECOND EDITION

I AM GRATEFUL to Basic Books for bringing out the second edition of *The Devil's Delusion: Atheism and its Scientific Pretensions* and grateful to those who made it possible: Lara Heimert, Susan Ginsburg, Diana Banister, Steven Meyer, Rob Crowther and John West.

I could have no better friends.

Apart from correcting a few typographic mistakes and pruning a few superfluous sentences from the text, I have made no changes to the original edition.

A T THE beginning of his *Letter to a Christian Nation,* Sam Harris writes that his fiercest and most "disturbed" critics are Christians who are "deeply, even murderously, intolerant of criticism." It would seem that a good many of those intolerant critics have been sending Harris biblical verses supporting their intolerance. Now, I count myself among Harris's warmest detractors. When he remarks that he has been *dumbstruck* by Christian and Moslem intellectual commitments, I believe the word has met the man. But here it is, an inconvenient fact: I am a secular Jew. My religious education did not take. I can barely remember a word of Hebrew. I cannot pray. I have spent more years than I care to remember in studying mathematics and writing about the sciences. Yet the book that follows is in some sense a defense of religious thought and sentiment. Biblical verses are the least of it.

A defense is needed because none has been forthcoming. The discussion has been ceded to men who regard religious belief with frivolous contempt. Their books have in recent years poured from every press, and although differing widely in their style, they are identical in their message: Because scientific theories are true, religious beliefs must be false. Harris has conveyed the point by entitling an essay "Science Must Destroy Religion." His call to jihad cannot be long delayed.

If science stands opposed to religion, it is not because of anything contained in either the premises or the conclusions of the great scientific theories. They do not mention a word about God. They do not treat of any faith beyond the one that they themselves demand. They compel no ritual beyond the usual rituals of academic life, and these involve nothing more than the worship of what is widely worshipped. Confident assertions by scientists that in the privacy of their chambers they have demonstrated that God does not exist have nothing to do with science, and even less to do with God's existence.

In all this, two influential ideas are at work. The first is that there is something answering to the name of science. The second is that something answering to the name of science offers sophisticated men and women a coherent vision of the universe. The second claim is false if the first claim is.

And the first claim *is* false. *Nothing* answers to the name of science. And Nothing has no particular method either, beyond the immemorial dictates of common sense.

Like *democracy* or *justice*, *science* is a word exhausted by its examples. We have been vouchsafed four powerful and profound scientific theories since the great scientific revolution of the West was set in motion in the seventeenth century—Newtonian mechanics, James Clerk Maxwell's theory of the electromagnetic field, special and general relativity, and quantum mechanics. These are isolated miracles, great mountain peaks surrounded by a range of low, furry foothills. The theories that we possess are "magnificent, profound, difficult, sometimes phenomenally accurate," as the distinguished mathematician Roger Penrose has observed, but, as he at once adds, they also comprise a "tantalizingly *inconsistent* scheme of things."

These splendid artifacts of the human imagination have made the world more mysterious than it ever was. We know better than we did what we do not know and have not grasped. We do not know how the universe began. We do not know why it is there. Charles Darwin talked speculatively of life emerging from a "warm little pond." The pond is gone. We have little idea how life emerged, and cannot with assurance say that it did. We cannot reconcile our understanding of the human mind with any trivial theory about the manner in which the brain functions. Beyond the trivial, we have no other theories. We can say nothing of interest about the human soul. We do not know what impels us to right conduct or where the form of the good is found.

On these and many other points as well, the great scientific theories have lapsed. The more sophisticated the theories, the more inadequate they are. This is a reason to cherish them. They have enlarged and not diminished our sense of the sublime.

No scientific theory touches on the mysteries that the religious tradition addresses. A man asking why his days are short and full of suffering is not disposed to turn to algebraic quantum field theory for the answer. The answers that prominent scientific figures *have* offered are remarkable in their shallowness. The hypothesis that we are nothing more than cosmic accidents has been widely accepted by the scientific community. Figures as diverse as Bertrand Russell, Jacques Monod, Steven Weinberg, and Richard Dawkins have said it is so. It is an article of their faith, one advanced with the confidence of men convinced that nature has equipped them to face realities the rest of us cannot bear to contemplate. There is not the slightest reason to think this so.

While science has nothing of value to say on the great and aching questions of life, death, love, and meaning, what the religious traditions of mankind *have* said forms a coherent body of thought. The yearnings of the human soul are not in vain. There is a system of belief adequate to the complexity of experience. There is recompense for suffering. A principle beyond selfishness is at work in the cosmos. All will be well.

I do not know whether any of this is true. I am certain that the scientific community does not know that it is false.

Occupied by their own concerns, a great many men and women have a dull, hurt, angry sense of being oppressed by the sciences. They are frustrated by endless scientific boasting. They suspect that as an institution, the scientific community holds them in contempt. They feel no little distaste for those speaking in its name.

They are right to feel this way. I have written this book for them.

CHAPTER

1

No Gods Before Me

U~NTIL JUST~ yesterday, it was fashionable for scientists carefully to cast their bread upon various ecclesiastical waters. *Very* carefully. In writing about Darwin's God, the biologist Kenneth Miller affirmed that he saw no conflict whatsoever between his own Catholic faith and Darwin's theory of evolution. Francis Collins, who directed the Human Genome Project, has made a very similar case for his religious beliefs. Science and religion, Stephen Jay Gould remarked, constitute Non-Overlapping Magisteria. Science is a fine thing. Religion is a fine thing too. They are two very fine things. The great master of this tolerant spirit was Albert Einstein. What was it he said? "Science without religion is lame, religion without science is blind." The lame and the blind excepted, who could object?

If scientists were unwilling to give offense to religion, perhaps from a decent sense that it was precisely their religious belief that enabled many men and women the better to endure life, they were very often equally unwilling enthusiastically to endorse its conclusions. And for the same underlying reason: *Why make trouble?* When the great Austrian logician Kurt Gödel devised an interesting version of the ontological argument, he showed it to friends and warned them that having created an argument in favor of God's existence, he was not about to believe his own conclusions. He had merely been testing the limits of his intellectual power. It is something, after all, that every man might wish to know.

With the rise of what the *Wall Street Journal* has called "militant atheism," both the terms of debate and the climate of opinion have changed. The sunny agnosticism characteristic of men who believed that with respect to God, it could go either way, is no longer in fashion. It is regarded as rather dim.

Some of this represents nothing more than the reappearance of that perennial literary character, the village atheist, someone prepared tediously to dispute the finer points of Second Corinthians in time taken from spring planting. A little philosophy, as Francis Bacon observed, "inclineth man's mind to atheism." A *very* little philosophy is often all that is needed. In a recent BBC program entitled *A Brief History of Unbelief,* the host, Jonathan Miller, and his guest, the philosopher Colin McGinn, engaged in a veritable orgy of competitive skepticism, so much so that in the end, the viewer was left wondering whether either man believed sincerely in the existence of the other. Sam Harris's *Letter to a Christian Nation* is in this tradition, and if his book is devoid of any intellectual substance whatsoever, it is, at least, brisk, engaging, and short. To anyone having read Daniel Dennett's *Breaking the Spell: Religion as a Natural Phenomenon*, these will appear as very considerable virtues.

If rural atheism is familiar, it is also irrelevant. Religious men and women, having long accommodated the village idiot, have long accommodated the village atheist. The order of battle is now different. It has been the *scientists*—Richard Dawkins, Victor Stenger, Taner Edis, Emile Zuckerkandl, Peter Atkins, Steven Weinberg (*vasta môle superbus*)—who have undertaken a wide-ranging attack on religious belief and sentiment. Although efforts among atheists to promote fellowship by calling one another *bright* have not, it must be said, proven a great success, in all other respects, their order is thriving. Richard Dawkins, the author of *The God Delusion*, is in this respect outstanding. He is not only an intellectually fulfilled atheist, he is determined that others should be as full as he. A great many scientists are satisfied that at last someone has said out loud what so many of them have said among themselves: *Scientific and religious belief are in conflict. They cannot both be right. Let us get rid of the one that is wrong.* Where before he was tolerated, Dawkins is now admired. Should he announce that shortly he will conduct a personal invasion of Hell in order to roust various American evangelicals, ticket sales at the National Academy of Sciences would at once start vibrating.

These views are important because they invoke for their authority the power and the glory of the Western scientific tradition. The title of Victor Stenger's recent book is *God: The Failed Hypothesis—How Science Shows That God Does Not Exist.* Stenger is a professor of physics. He may have written the book, but it is *science*, we are to understand, that has provided the requisite demonstration. Like a nineteenth-century spirit medium, Stenger has simply taken dictation. The physicist Taner Edis has also seen the light, and so published a book. Entitled *The Ghost in the Universe,* it is not a celebration of the Host. Both men exhibit the salient characteristic of physicists endeavoring to draw general lessons about the cosmos from mathematical physics: They are willing to believe in anything.

Because atheism is said to follow from various scientific doctrines, literary atheists, while they are eager to speak their minds, must often express themselves in other men's voices. Christopher Hitchens is an example. With forthcoming modesty, he has affirmed his willingness to defer to the world's "smart scientists" on any matter more exigent than finger-counting. Were smart scientists to report that a strain of yeast supported the invasion of Iraq, Hitchens would, no doubt, conceive an increased respect for yeast. He is presently persuaded that "religion poisons everything." His book is entitled *God Is Not Great,* and within its pages, he has managed to convey his contempt for religious thought by propositions exhibiting a positively oriental degree of evasiveness. "We do not rely solely upon science and reason," he writes, "because these are necessary rather than sufficient factors, but we distrust anything that contradicts science or outrages reason." If Hitchens is not prepared to "rely solely upon science and reason," why, one might ask, should anyone else? And if science and reason are "necessary rather than sufficient factors," then who is to say that factors both necessary *and* sufficient might not convey a man to the very edge of faith? It is by means of these questions, I imagine, that one day the lion shall lie down with the lamb, circumstances that with justifiable pride Hitchens may affirm that he has anticipated.

☙

Does any of this represent anything more than yet another foolish intellectual fad, a successor to academic Marxism, feminism, or various doctrines of multicultural tranquillity? Not in the world in which religious beliefs overflow into action. For Islamic radicals, "the sword is more telling than the book," as the Arab poet Abu Tammam wrote with menacing authority some eight hundred years ago. The advent of militant atheism marks a reaction—a lurid but *natural* reaction—to the violence of the Islamic world.

But the efflorescence of atheism involves more than atheism itself. Of course it does. Atheism is the *schwerpunkt*, as German military theorists used to say with satisfaction, the place where force is concentrated and applied; and what lies behind is a doctrinal system, a way of looking at the world, and so an ideology. It is an ideology with no truly distinct center and the fuzziest of boundaries. For purposes of propaganda it hardly matters. Science as an institution is unified by the lowest common denominator of belief, and that is the conviction that science is a very good thing.

Curiously enough, for all that science may be a very good thing, members of the scientific community are often dismayed to discover that, like policemen, they are not better loved. Indeed, they are widely considered self-righteous, vain, politically immature, and arrogant. This last is considered a special injustice. "Contrary to what many anti-intellectuals maintain," the biologist Massimo Pigliucci has written, science is "a much more humble enterprise than any religion or other ideology." Yet despite the *outstanding* humility of the scientific community, anti-intellectuals persist in their sullen suspicions. Scientists are hardly helped when one of their champions immerses himself in the emollient of his own enthusiasm. Thus Richard Dawkins recounts the story of his professor of zoology at Oxford, a man who had "for years . . . passionately believed that the Golgi apparatus was not real." On hearing during a lecture by a visiting American that his views were in error, "he strode to the front of the hall, shook the American by the hand, and said—with passion—'My dear fellow, I wish to thank you. I have been wrong these fifteen years.' " The story, Dawkins avows, still has the power "to bring a lump to my throat."

It could not have been a very considerable lump. No similar story has ever been recounted about Richard Dawkins. Quite the contrary. He is as responsive to criticism as a black hole in space. "It is absolutely safe to say," he has remarked, "that if you meet somebody who claims not to believe in evolution that person is ignorant, stupid or insane."

The tone is characteristic. Peter Atkins is a professor of physical chemistry at Oxford University, and he, too, is ardent in his atheism. In the course of an essay denouncing not only theology but poetry and philosophy as well, he observes favorably of himself that scientists "are at the summit of knowledge, beacons of rationality, and intellectually honest." It goes without saying, Atkins adds, that "there is no reason to suppose that science cannot deal with every aspect of existence." Science is, after all, "the apotheosis of the intellect and the consummation of the Renaissance."

These comical declarations may be abbreviated by observing that Atkins is persuaded that not only is science a very good thing, but no other thing is good at all.

᪤

Ever since the great scientific revolution was set in motion by Johannes Kepler, Galileo Galilei, and Isaac Newton, it has been a commonplace of commentary that the more that science teaches us about the natural world, the less important a role human beings play in the grand scheme of things. "Astronomical observations continue to demonstrate," Victor Stenger affirms, "that the earth is no more significant than a single grain of sand on a vast beach." What astronomical observations may, in fact, have demonstrated is that the earth is no more *numerous* than a single grain of sand on a vast beach. *Significance* is, of course, otherwise. Nonetheless, the inference is plain: What holds for the earth holds as well for human beings. They hardly count, and scientists like Stenger are not disposed to count them at all. It is, as science writer Tom Bethell notes, "an article of our secular faith that there is nothing exceptional about human life."

The thesis that we are all nothing more than vehicles for a number of "selfish genes" has accordingly entered deeply into the simian gabble of academic life, where together with materialism and moral relativism it now seems as self-evident as the law of affirmative action. To anyone who has enjoyed the spectacle of various smarmy insects shuffling along the tenure track at Harvard or Stanford, the idea that we are all simply "survival machines" seems oddly in conflict with the correlative doctrine of the survival of the fittest. This would not be the first time that an ideological system in conflict with the facts has found it prudent to defer to itself.

And with predictably incoherent results. After comparing more than two thousand DNA samples, an American molecular geneticist, Dean Hamer, concluded that a person's capacity to believe in God is linked to his brain chemicals. Of all things! Why not his urine? Perhaps it will not be amiss to observe that Dr. Hamer has made the same claim about homosexuality, and if he has refrained from arguing that a person's capacity to believe in molecular genetics is linked to a brain chemical, it is, no doubt, owing to a prudent sense that once *that* door is open, God knows how and when anyone will ever slam it shut again.

Neither scientific credibility nor sound good sense is at issue in *any* of these declarations. They are absurd; they are understood to be absurd; and what is more, assent is demanded just *because* they are absurd. "We take the side of science *in spite* of the patent absurdity of some of its constructs," the geneticist Richard Lewontin remarked equably in *The New York Review of Books*, "*in spite* of its failure to fulfill many of its extravagant promises of health and life, *in spite* of the tolerance of the scientific community for unsubstantiated just-so stories" (my emphasis).

Why should any discerning man or woman take the side of science, or anything else, under these circumstances? It is because, Lewontin explains, "we cannot allow a Divine Foot in the door."

If one is obliged to accept absurdities for fear of a Divine Foot, imagine what prodigies of effort would be required were the rest of the Divine Torso found wedged at the door and with some justifiable irritation demanding to be let in?

If nothing else, the attack on traditional religious thought marks the consolidation in our time of science as the single system of belief in which rational men and women might place their faith, and if not their faith, then certainly their devotion. From cosmology to biology, its narratives have become *the* narratives. They are, these narratives, immensely seductive, so much so that looking at them with innocent eyes requires a very deliberate act. And like any militant church, this one places a familiar demand before all others: Thou shalt have no other gods before me.

It is this that is new; it is this that is important.

2

Nights of Doubt

WHETHER GOD exists—that is one question. Whether belief in his existence plays an important role in human life—that is another. "Religion's power to console," Richard Dawkins writes in *The God Delusion,* "doesn't make it true." Perhaps this is so, but only a man who has spent a good deal of time snoring on the down of plenty could be quite so indifferent to the consolations of religion, wherever and however they may be found. One wonders, in any case, why religion *has* the power to console and why it has had this power over the course of human history.

Writing about the arts and their degraded state, Camille Paglia begins by affirming that she is a "professed atheist."

She is nonetheless persuaded that "a totally secularized society with contempt for religion sinks into materialism and self-absorption and gradually goes slack." The connection between what she sees (a good deal that is awful) and what she believes (There is no God) is not one that she is inclined to make. When faced with irreconcilable alternatives, she proposes to straddle the difference, a position as difficult in thought as it is uncomfortable in gymnastics. Her calls for the study of comparative religion at least afford the consumer the luxury of choice without the penalty of commitment. "I view each world religion," she writes, "as a complex symbol system, a metaphysical lens through which we can see the vastness and sublimity of the universe."

I daresay that a telescope does a better job in revealing the size of the universe than any of the world religions, and if sublimity is wanted, it is hardly to be expected from a system of thought assumed to be *false*.

There remains another possibility. There may in fact be a connection between the importance of religious belief in life and the existence of the Deity in reality.

Not a logical connection, no. But a connection nonetheless, and so a clue.

And let us be honest: When it comes to clues, we could all use a few more.

THE SEEING EYE

During the living centuries of the Arab empire, a magnificent series of stellar observatories glittered like jewels throughout the archipelago of its conquests. The observatory played an important role in the religious life of devout Moslems. It was not—*it was never*—the expression of disinterested curiosity. More so than either Jews or Christians, men of the Moslem faith were called upon carefully to mark the schedule of their devotions. The art devoted to such concerns was known as the *ilm al-miqât*. And an art it was. During the Middle Ages, the Moslem world, for all its luxury and sophistication, had no more access to sophisticated clocks than the Christian world, and in the Christian West, men kept time so carelessly that even the arrival of the Easter holidays was a matter of profound uncertainty. Caliphs in Baghdad counted time by means of either a water clock or an hourglass, and yet the Koran commanded fivefold prayers each day, and it commanded the faithful to face the shrine of Kaaba in Mecca as they prayed— tasks requiring considerable mental dexterity. The Islamic calendar was based on the phases of the moon. The community preparing to celebrate the holy month of Ramadan, which marks the beginning of the lunar year, would need to spot the crescent moon just as it shed its blush in the evening sky. Before the creation of sophisticated astronomical tables, men with exceptionally sharp eyesight were sent to distant mountaintops to spot the moon's appearance; their cries then echoed down through the valleys and thence by a chain of cries back to Baghdad itself. (In France, the night of the crescent moon is still called *la nuit de doute*—the night of doubt.) By the thirteenth century, these scientific chores were assigned to professionals, the so-called *muwaqqit*. Resident in mosques, they were responsible for regulating the time of prayer. "In Islam, as in no other religion," the historian David King has remarked, "the performance of various aspects of religious ritual has been assisted by scientific procedure."

A BESTIAL INDULGENCE OF APPETITE

And now a question: Does the Koran commend the study of the natural world? And an answer: It does. "At the last Judgment," the Turkish devout Said Nursî remarked, "the ink spent by scholars is equal to the blood of martyrs." But those scholars celebrated at the last judgment were apt to be scholars of religion and so bound by the inerrancy of the Koran. "Allah turns over the night and the day," reads a well-known Koranic verse, "most surely there is a lesson in this for those who have sight" (24.44). It is hardly surprising that Moslem mathematicians and astronomers, from the late seventh to the early fifteenth century, regarded their scientific curiosity, on those occasions when they were called upon to justify it, as if their scientific pursuits comprised an exercise calculated to increase their devotion.

But of all the human emotions, curiosity is the one least subject to the general proscription against gluttony, and once engaged, even if engaged initially in the service of religion, it has a tendency to grow relentlessly, until in the end the scholar becomes curious about the nature of revelation itself. The more encompassing the scope of scholarship, the more open to doubt the scholar becomes, so that in the end only curiosity remains indisputably of value. This is true whether the object of curiosity is religion *or* science.

Writing in 1420 or 1430, the astronomer Ulugh Beg described science in a way that suggests nothing of the martyr's blood. "Intellects are in agreement," he wrote, "and minds are in accord as to the excellence of science and the worthiness of scientists." By "science," Ulugh Beg meant observation—the power of the eye, aided by various instruments, to *see*. The benefits conferred by sight are very often matters of self-improvement. "Science sharpens the intellect and strengthens it; it increases sagacity, and augments perspicacity." But benefits transcend the personal. Those sciences whose principles are "indisputable and self-evident" have the merit of being "common to people of different religions," Ulugh Beg affirmed.

These sentiments are entirely modern. They might well have been expressed by a committee of the National Science Foundation. They *were* expressed by a committee of the National Science Foundation: "Science extends and enriches our lives, expands our imagination and liberates us from the bonds of ignorance and superstition." They are on display in every high school textbook.

And there is hardly any reason to suppose them true.

It is a point that did not fail to escape the notice of the most perceptive of the Arab philosophers, the *gazelle*, Abu Hamid Muhammad Al-Ghazâli. Writing with remarkable prescience about the scientists he called *naturalists*, and this in the eleventh century, Al Ghazâli was quite prepared to admit that their studies served to reveal "the wonders of creation." No one "can make a careful study of anatomy and the wonderful uses of the members and organs [of the human body] without attaining to the necessary knowledge that there is a perfection in the order which the framer gave to the animal frame, and especially to that of man."

At once, Al Ghazâli withdraws the commendation that he has just offered. A complicated inference is set in play. The naturalists argue, he observes, that "intellectual power in man is dependent on [his] temperament." It is a point that neurophysiologists would today make by arguing that the mind (or the soul) is dependent on the brain, or even that the mind *is* the brain. From this it follows that "as the temperament is corrupted, *intellect is also corrupted and ceases to exist.*" When the brain is destroyed, so, too, the mind. Death and disease mark the end of the mind. On the naturalistic view, Al Ghazâli argues, "the soul dies and does not return to life." The globe of consciousness shrinks in each of us until it is no larger than a luminous point, and then it winks out.

But if this is a matter of fact, Al Ghazâli argues, it is a matter of profound scientific *and* moral consequence. Why should a limited and finite organ such as the human brain have the power to see into the heart of matter or mathematics? These are subjects that have nothing to do with the Darwinian business of scrabbling up the greasy pole of life. It is as if the liver, in addition to producing bile, were to demonstrate an unexpected ability to play the violin. This is a question that Darwinian biology has not yet answered. By the same token, to place in doubt the survival of the soul is to "deny the future life—heaven, hell, resurrection, and judgment." And this is to corrupt the system of justice by which life must be regulated, because "there does not remain any reward for obedience, or any punishment for sin."

With this curb removed, Al Ghazâli predicts, men and women will give way to "a bestial indulgence of their appetites."

As he so often does, Al Ghazâli has managed to express a very complex current of anxiety common not only in the Moslem world but in the world at large.

⊷

If it is hardly unknown, this medieval Arabic anxiety, it no longer controls the moral imagination in any secular society. It does not control *mine* and I suppose it does not control yours either. A great many men and women do suspect that scientific curiosity, if unchecked, might be a dangerous force. Like any dangerous force, scientific curiosity is dangerous because in the end it turns upon itself. The stories both of Faust and Frankenstein suggest that this is so. But a bestial indulgence of appetite? This is not a phrase, nor does it evoke an idea, that anyone in the West now finds plausible. Quite the contrary. It is *religion*, Christopher Hitchens claims, that is dangerous, because it is "the cause of dangerous sexual repression." Short of gender insensitivity, what could be more dangerous than dangerous sexual repression? Among the commandments that Richard Dawkins proposes as replacements for the original ten, the first encourages men and women "to enjoy [their] own sex lives so long as it damages nobody else." What Hector Avalos has called "the Enlightenment project" of allowing men and women to regulate their own conduct by means "reason and experience" may in the early twenty-first century have led to a certain tastelessness in public entertainment, but what of it?

Worse things have happened.

The conviction that in Western Europe and the United States nothing worse *has* happened is one reason that so many scientific atheists affirm that they are of the Enlightenment party. It is a party everyone is eager to join, Noam Chomsky because he is a "child" of the Enlightenment, the rest of us because for the moment, there are no other parties at all.

Children of the Enlightenment do not, of course, dwell overly on the dreadful acts undertaken in its name when the Enlightenment first became a living historical force in France: *all perished, all— / Friends, enemies, of all parties, ages, ranks, / Head after head, and never heads enough / For those that bade them fall.*

Why should the sins of the fathers be visited on their children?

DOUBLE-ENTRY BOOKKEEPING

For scientists persuaded that there is no God, there is no finer pleasure than recounting the history of religious brutality and persecution. Sam Harris is in this regard especially enthusiastic, *The End of Faith* recounting in lurid but lingering detail the methods of torture used in the Spanish Inquisition. If readers require pertinent information concerning the strappado, or other instruments of doctrinal persuasion, they may turn to his pages. There is no need to argue the point. A great deal of human suffering has been caused by religious fanaticism. If the Inquisition no longer has the power to compel our indignation, the Moslem world often seems quite prepared to carry the burden of exuberant depravity in its place.

Nonetheless, there is this awkward fact: The twentieth century was not an age of faith, and it was awful. Lenin, Stalin, Hitler, Mao, and Pol Pot will never be counted among the religious leaders of mankind.

Nor can anyone argue that the horrors of the twentieth century were unanticipated. Although they came as a shock, they did not come as a surprise. In *The Brothers Karamazov*, Ivan Karamazov exclaims that if God does not exist, then everything is permitted. Throughout the nineteenth century, as religious conviction seeped out of the institutions of Western culture, poets and philosophers had the uneasy feeling that its withdrawal might signal the ascension of great evil in the world.

In this they were right.

What gives Karamazov's warning—for that is what it is—its power is just that it has become part of a most up-to-date hypothetical syllogism:

The first premise:

If God does not exist, then everything is permitted.

And the second:

If science is true, then God does not exist.

The conclusion:

If science is true, then everything is permitted.

Whereupon there is a return to a much older, vastly more somber vision of life and its constraints, one that serves to endow the phrase *bestial indulgence* with something more by way of content than popularly imagined.

᠖

In 2007, a number of scientists gathered in a conference entitled "Beyond Belief: Science, Religion, Reason, and Survival" in order to attack religious thought and congratulate one another on their fearlessness in so doing. The physicist Steven Weinberg delivered an address. As one of the authors of the theory of electroweak unification, the work for which he was awarded a Nobel Prize, he is a figure of great stature. "Religion," he affirmed, "is an insult to human dignity. With or without it you would have good people doing good things and evil people doing evil things. *But for good people to do evil things, that takes religion*" (italics added).

In speaking thus, Weinberg was warmly applauded, not one member of his audience asking the question one might have thought pertinent: Just *who* has imposed on the suffering human race poison gas, barbed wire, high explosives, experiments in eugenics, the formula for Zyklon B, heavy artillery, pseudo-scientific justifications for mass murder, cluster bombs, attack submarines, napalm, intercontinental ballistic missiles, military space platforms, and nuclear weapons?

If memory serves, it was not the Vatican.

❧

If the facts about the twentieth century are an inconvenience for scientific atheism, suitably informed thought may always find a way to deny them. The psychologist Steven Pinker has thus introduced into the discussion the remarkable claim that "something in modernity and its cultural institutions has made us nobler."

The good news is unrelenting: "On the scale of decades, comprehensive data again paint a shockingly happy picture."

"Some of the evidence," Pinker goes on to say, "has been

under our nose all along. Conventional history has long shown that, in many ways, we have been getting kinder and gentler.

Cruelty as entertainment, human sacrifice to indulge superstition, slavery as a labor-saving device, conquest as the mission statement of government, genocide as a means of acquiring real estate, torture and mutilation as routine punishment, the death penalty for misdemeanors and differences of opinion, assassination as the mechanism of political succession, rape as the spoils of war, pogroms as outlets for frustration, homicide as the major form of conflict resolution—all were unexceptionable features of life for most of human history. But, today, they are rare to nonexistent in the West, far less common elsewhere than they used to be, concealed when they do occur, and widely condemned when they are brought to light.

Here is rather a more accurate assessment of the twentieth and early twenty-first centuries. Anyone persuaded that they represent a "shockingly happy picture" should make the modest imaginative effort to discern the immense weight of human misery conveyed by these statistics:

A Shockingly Happy Picture by Excess Deaths

First World War (1914–18):	15 million
Russian Civil War (1917–22):	9 million
Soviet Union, Stalin's regime (1924–53):	20 million
Second World War (1937–45):	55 million
Chinese Civil War (1945–49):	2.5 million

People's Republic of China, Mao Zedong's
    regime (1949–75): .... 40 million
Tibet (1950 et seq.): 600,000
Congo Free State (1886–1908): .... 8 million
Mexico (1910–20): .... 1 million
Turkish massacres of Armenians (1915–23): ..... 1.5 million
China (1917–28): 800,000
China, Nationalist era (1928–37): .... 3.1 million
Korean War (1950–53): .... 2.8 million
North Korea (1948 et seq.): .... 2 million
Rwanda and Burundi (1959–95): .... 1.35 million
Second Indochina War (1960–75): .... 3.5 million
Ethiopia (1962–92): 400,000
Nigeria (1966–70): .... 1 million
Bangladesh (1971): .... 1.25 million
Cambodia, Khmer Rouge (1975–78): .... 1.65 million
Mozambique (1975–92): .... 1 million
Afghanistan (1979–2001): .... 1.8 million
Iran-Iraq War (1980–88): .... 1 million
Sudan (1983 et seq.): .... 1.9 million
Kinshasa, Congo (1998 et seq.): .... 3.8 million
Philippines Insurgency (1899–1902): 220,000
Brazil (1900 et seq.): 500,000
Amazonia (1900–1912): 250,000
Portuguese colonies (1900–1925): 325,000
French colonies (1900–1940): 200,000
Japanese War (1904–5): 130,000
German East Africa (1905–7): 175,000
Libya (1911–31): 125,000
Balkan Wars (1912–13): 140,000
Greco-Turkish War (1919–22): 250,000

A Shockingly Happy Picture by Excess Deaths *(cont'd)*

Spanish Civil War (1936-39): 365,000

Franco Regime (1939-75): 100,000

Abyssinian Conquest (1935-41): 400,000

Finnish War (1939-40): 150,000

Greek Civil War (1943-49): 158,000

Yugoslavia, Tito's regime (1944-80): 200,000

First Indochina War (1945-54): 400,000

Colombia (1946-58): 200,000

India (1947): 500,000

Romania (1948-89): 150,000

Burma/Myanmar (1948 et seq.): 130,000

Algeria (1954-62): 537,000

Sudan (1955-72): 500,000

Guatemala (1960-96): 200,000

Indonesia (1965-66): 400,000

Uganda, Idi Amin's regime (1972-79): 300,000

Vietnam, postwar Communist regime

   (1975 et seq.): 430,000

Angola (1975-2002): 550,000

East Timor, conquest by Indonesia (1975-99): 200,000

Lebanon (1975-90): 150,000

Cambodian Civil War (1978-91): 225,000

Iraq, Saddam Hussein (1979-2003): 300,000

Uganda (1979-86): 300,000

Kurdistan (1980s, 1990s): 300,000

Liberia (1989-97): 150,000

Iraq (1990-): 350,000

Bosnia and Herzegovina (1992-95): 175,000

Somalia (1991 et seq.): 400,000

In considering Pinker's assessment of the times in which we live, the only conclusion one can profitably draw is that such an excess of stupidity is not often to be found in nature.

AN INSULT TO HUMAN DIGNITY

Does something in the very nature of a secular society make the monstrous possible? At the very least, Hitler and Stalin would seem to offer the prosecution a good deal of space in which to maneuver.

And the defense?

Richard Dawkins accepts Stalin as a frank atheist, and so a liability of the sort that every family admits, but he is at least sympathetic to the thesis that Hitler's religious sentiments as a Catholic were sincere. Why stop with Hitler? No doubt some members of the SS took communion after an especially arduous day in the field murdering elderly Jewish women, and with vengeful Russian armies approaching Berlin, Heinrich Himmler, who had presided over the Third Reich's machinery of extermination and had supervised the desecration of churches and synagogues from one end of Europe to the other, confessed to an associate that he was persuaded of the existence of a Higher Power. The death of Franklin Roosevelt inspired Joseph Goebbels to similarly pious sentiments. The deathbed conversion is generally regarded as the mark of desperate insincerity. Throughout their careers, these scum acted as if no power was higher than their own. Dawkins is prepared to acknowledge the facts while denying their significance. Neither the Nazis nor the Communists, he affirms, acted *because* of their atheism. They were simply keen to kill a great many people. Atheism had nothing to do with it. They might well have been Christian Scientists.

In the early days of the German advance into Eastern Europe, before the possibility of Soviet retribution even entered their untroubled imagination, Nazi extermination squads would sweep into villages, and after forcing villagers to dig their own graves, murder their victims with machine guns. On one such occasion somewhere in Eastern Europe, an SS officer watched languidly, his machine gun cradled, as an elderly and bearded Hasidic Jew laboriously dug what he knew to be his grave.

Standing up straight, he addressed his executioner. "God is watching what you are doing," he said.

And then he was shot dead.

What Hitler did *not* believe and what Stalin did *not* believe and what Mao did *not* believe and what the SS did *not* believe and what the Gestapo did *not* believe and what the NKVD did *not* believe and what the commissars, functionaries, swaggering executioners, Nazi doctors, Communist Party theoreticians, intellectuals, Brown Shirts, Black Shirts, gauleiters, and a thousand party hacks did *not* believe was that God was watching what they were doing.

And as far as we can tell, very few of those carrying out

the horrors of the twentieth century worried overmuch that God was watching what they were doing either.

That is, after all, the *meaning* of a secular society.

⹂

One might think that in the dark panorama of wickedness, the Holocaust would above all other events give the scientific atheist pause. Hitler's Germany was a technologically sophisticated secular society, and Nazism itself, as party propagandists never tired of stressing, was "motivated by an ethic that prided itself on being scientific." The words are those of the historian Richard Weikart, who in his admirable treatise, *From Darwin to Hitler: Evolutionary Ethics, Eugenics, and Racism in Germany*, makes clear what anyone capable of reading the German sources already knew: A sinister current of influence ran from Darwin's theory of evolution to Hitler's policy of extermination. A generation of German biologists had read Darwin and concluded that competition between species was reflected in human affairs by competition between races.

These observations find no echo at all in the literature of scientific atheism. Christopher Hitchens is prepared to denounce the Vatican for the ease with which it diplomatically accommodated Hitler, but about Hitler, the Holocaust, or the Nazis themselves he has nothing to say. This is an odd omission for a writer who believes that *religion* poisons everything, and suggests that his eye for poison in political affairs tends under conditions of polemical stress to wander irresolutely.

When it comes to the Holocaust, Sam Harris, like so many others, approaches anti-Semitism and finds it surprisingly to his taste.

So far as the persecution of the Jewish people goes, Harris is opposed, if only because everyone outside the Arab world is. No great moral effort is required to reach this judgment.

"The gravity of Jewish suffering over the ages culminating in the Holocaust," Harris writes, "makes it almost impossible to entertain any suggestion that Jews might have brought their troubles on themselves."

Having rejected the suggestion as an impossibility, Harris at once proceeds to embrace it.

The Jewish people, it would appear, *did* bring their suffering on themselves for "their refusal to assimilate, for the insularity and professed superiority of their religious culture—*that is, for the content of their own sectarian beliefs*" (my emphasis). This is nicely in line with opinions advanced recently by the historian David Irving. "The Jews," he has concluded, "were the authors of their own misfortune." Recently released from an Austrian prison, where he had been incarcerated on charges of Holocaust denial, David Irving, like Typhoid Mary, is not generally considered a figure serious men and women are eager to enlist in their cause.

Although Harris is officially committed to assigning the blame for intolerance on the intolerant, there is blame enough left over to assign some to the intoleree as well. "The ideology of Judaism remains a lightning rod of intolerance *to this day*" (italics added). To be a lightning rod for intolerance is a moral defect, the more so when the remedy—get rid of those divisive sectarian beliefs—lies close at hand.

If you find it difficult to imagine that after close study of the Bava Mezia, the chapter of the Talmud that deals with the law of gifts, Hermann Göring decided that it was "the *ideology* of Judaism" that justified Nazi policies, then you have insufficiently appreciated just how divisive Jewish beliefs must have seemed to stout Göring, a man of well-known sensitivity to the delicacy of ideological deviance.

The contrary case has all the merits of the truth. For reasons that they could not make clear, *even to themselves,* the men controlling the Third Reich determined that it would be a fine thing to exterminate 9 million European Jews. In the SS and the German army, they found a willing instrument to hand. Much occupied in the closing days of the war with preserving their reputation—their reputation for diabolical wickedness—members of the SS took a perverse satisfaction in assuring one another that whatever they had done, it would not be believed, and if believed, blame would be assigned to their victims. In this, they were correct. More than fifty years after the Holocaust, a great many placid, well-meaning, and well-fed men and women persist in imagining that however monstrous the Holocaust, deep down the Jewish people, if they did not encourage their destruction, invited it nonetheless. "Judaism is as intrinsically divisive, as ridiculous in its literalism, and at odds with the civilizing insights of modernity as any other religion."

No doubt the civilizing insights of modernity appear considerable in Santa Barbara, where Sam Harris lives; but as travel broadens one's mind, it enlarges one's perspective, and those civilizing insights of which he writes are apt to seem a good deal less persuasive five thousand miles farther to the east, where modernity expressed itself in cattle cars rumbling from all the ancient civilized cities of Europe in order days later to deposit their famished, suffering victims at German extermination camps.

Some insight. Some modernity. Some civilization.

Having dismissed Jewish beliefs as divisive, Harris is concerned to affirm that they are *misguided* as well:

"It appears that even the Holocaust did not lead most Jews to doubt the existence of an omnipotent and benevolent God. If having half of your people delivered to the furnace does not count as evidence against the notion that an all-powerful God is looking out for your interests, it seems reasonable to assume that nothing could."

On the other hand, I suppose that Harris might speculate on what is equally an interesting matter of *evidence*, a concept that he values in the abstract and on every occasion ignores in the particular. The Jewish people yet live, and even in Eastern Europe—even in Poland—they have returned to their ancestral homes; but the thousand-year Reich, *that* lies buried in the rubble of German cities smashed to smithereens, or ground under Russian tank treads, or destroyed by American artillery, or left to wander in its exiled millions across all the violated borders of Central Europe, and if God did not protect his chosen people precisely as Harris might have wished, He did, in an access of his old accustomed vigor, smite their enemies, with generations to come in mourning or obsessed by shame.

TROUBLINGLY UNTROUBLED

There is a queer quality of logical brittleness to everything that Harris writes, because every argument he advances stops before it has become relevant. The moral concerns that are prompted by biology? The list is already long: abortion, stem-cell research, euthanasia, infanticide, cloning, animal-human hybrids, sexual deviancy. It will get longer, as scientists with no discernible sense of responsibility to human nature come extravagantly to interfere in human life. In his *Letter to a Christian Nation*, Harris argues that "qualms" about stem-cell research are "obscene," because they are "morally indefensible." And they are morally indefensible because they represent nothing more than "faith-based irrationality."

These remarks are typical; they embody a style. And they invite the obvious response. Beyond the fact that it is religiously based, just what makes the religious objection to stem-cell research irrational?

Those who find these questions troubling—me, for sure— find them troubling because atheists such as Sam Harris remain so resolutely untroubled by them. His convictions are as tranquil as his face is unlined. That bat squeak of warning that so many religious believers hear when they consider stem-cell research, abortion, or euthanasia sounds at a frequency to which he is insensitive.

This is very odd considering that what moral philosophers have called the slippery slope has proven in recent decades to be slippery enough to seem waxed. It is, if anything, more slippery than ever. In 1984, Holland legalized euthanasia. Critics immediately objected that Dutch doctors, having been given the right to kill their elderly patients at their request, would almost at once find reasons to kill patients at their whim. This is precisely what has happened. The *Journal of Medical Ethics,* in reviewing Dutch hospital practices, reported that 3 percent of Dutch deaths for 1995 were assisted suicides, and that of these, fully one-fourth were involuntary. The doctors simply knocked their patients off, no doubt assuring the family that *Grootmoeder* would have wanted it that way. As a result, a great many elderly Dutch carry around sanctuary certificates indicating in no uncertain terms that they do not wish their doctors to assist them to die, emerging from their coma, when they are ill, just long enough to tell these murderous pests for heaven's sake to go away. The authors of the study, Henk Jochensen and John Keown, reported with some understatement that "Dutch claims of effective regulation ring hollow."

Euthanasia, as Dr. Peggy Norris observed with some asperity, "cannot be controlled."

If this is so, why is Harris so sure that stem-cell research can be controlled?

And if it cannot be controlled, just what is irrational about religious objections to social policies that when they reach the bottom of the slippery slope are bound to embody something Dutch, degraded, and disgusting?

How many scientific atheists, I wonder, propose to spend their old age in Holland?

WHAT MAKES MEN GOOD?

Nothing. This is the answer of historical experience and a troubled common sense. It is the answer of Christian theology, and finds its expression in the doctrine of original sin. Having been asked by his biographer, James Boswell, for his opinion of original sin, Dr. Johnson responded in words to which he drew particular attention: "With respect to original sin, *the inquiry is not necessary,* for whatever is the cause of human corruption, men are evidently and confessedly so corrupt, that all the laws of heaven and earth are insufficient to restrain them from crimes" (italics added).

One need hardly be a Christian to appreciate the wisdom in these remarks. When Christopher Hitchens asks how much self-respect "must be sacrificed in order that one may squirm continually in an awareness of one's own sin," the only honest answer is that for most of us, self-respect is possible only if the squirming is considerable.

Men are not by nature good. Quite often, quite the contrary. And for this reason they must be restrained, by threats if possible, by force if necessary. "Perhaps," Richard Dawkins speculates, "I . . . am a Pollyanna to believe that people would remain good when unobserved and unpoliced by God."

I am under most circumstances the last person on earth to think Richard Dawkins a Pollyanna, but in this case I defer to his description. Why *should* people remain good when unobserved and unpoliced by God? Do people remain good when unpoliced by the police? If Dawkins believes that they do, he must explain the existence of the criminal law, and if he believes that they do not, then he must explain why *moral* enforcement is not needed at the place where law enforcement ends.

To scientific atheists, the ancient idea that *homo homini lupus*—man is a wolf to man—leaves them shaking their heads in poodle-like perplexity. Sam Harris has no anxieties whatsoever about presenting his own views on human morality with the enviable confidence of a man who feels that he has reached the epistemological bottom. "Everything about human experience," he writes, "suggests that love is more conducive to human happiness than hate is." It goes without saying, of course, that Harris believes that this is an *objective* claim about the human mind.

If this is so, it is astonishing with what eagerness men have traditionally fled happiness.

THE STILL SMALL VOICE

I f the universe is as scientists say it is, then what scope remains for statements about right or wrong, good or bad? What are we to say about evil and great wickedness? Whatever statements we might make are obviously not about gluons, muons, or curved space and time. "The problem," the philosopher Simon Blackburn has written, "is one of finding room for ethics, or of placing ethics within the disenchanted, nonethical order which we inhabit, and of which we are a part."

Blackburn is, of course, convinced that the chief task at hand in facing this question—his chief task, in any case—"is above all to refuse appeal to a supernatural order." It is a strategy that merits admiration for the severity of mind it expresses. It is rather as if an accomplished horseman were to decide that his chief task were to learn to ride without a horse.

If moral statements are about something, then the universe is not quite as science suggests it is, since physical theories, having said nothing about God, say nothing about right or wrong, good or bad. To admit this would force philosophers to confront the possibility that the physical sciences offer a grossly inadequate view of reality. And since philosophers very much wish to think of themselves as scientists, this would offer them an unattractive choice between changing their allegiances or accepting their irrelevance.

These are familiar questions in philosophy, and if they have been long asked, they have remained long unanswered. David

Hume asked in the eighteenth century whether *ought* could be derived from *is,* and concluded that it could not: There is a gap between what is and what ought to be. The world of fact and the world of value are disjoint. They have nothing to say to one another. The ensuing chilliness between what is and what ought to be has in the twentieth century grown glacial. The more that science reveals what is, the less it reveals what ought to be. The traditional biblical view—that what ought to be is a matter chiefly of what God demands—thus stands on his existence, the very point challenged by scientific atheism.

But if scientific atheists are disposed to challenge God's existence—the party line, after all—they are far less willing to reflect on what His dismissal entails. At some time after it had become clear that Nazi Germany would lose the Second World War, and before the war had actually been lost, one of the senior party officers—perhaps it was Himmler—in confronting the very complicated series of treaty obligations that Germany had accepted with respect to its satraps, wondered out loud, "What, after all, *compels* us to keep our promises?" It is a troubling question, and one that illustrates anew the remarkable genius for moral philosophy that the Nazis enjoyed.

What does?

ϛᢔ

In many ways, the issues raised by the existence of moral laws suggest a surprising connection between the laws of physics and the laws of morality. In both cases, questions arise very quickly as to the source of such laws and the reason for their truth.

We do not know why the laws of nature are true, even though we can sense that the question hides some sort of profound mystery.

A similar discussion has long been current in philosophy and has its source in Plato's *Euthyphro*. There Socrates asks whether what is good is good because the gods have declared it so, or whether the gods have declared it so because it is good.

To the question what makes the laws of moral life true, there are three answers: God, logic, and nothing. Each is inadequate.

If moral laws reflect the will of God, then He might presumably change his mind, and tomorrow issue a new set of commandments encouraging rape, plunder, murder, or the worship of false idols. Many devoutly religious men and women would say that this is his perfect right. He is God, after all. But if tomorrow God were to encourage rape as a very good thing, would rape *become* a very good thing, or would we conclude, along with Richard Dawkins, that considering his poor life choices, God is a repellent figure and to hell with Him?

If, on the other hand, God chooses the right or the good because it is right or good, then the power of his imperative has its source in the law, and not in his will. "Thou shall not kill," we may imagine God saying to the ancient Hebrews, "because it is wrong. I am here only to convey the message."

If this is so, then God must be demoted to what is plainly a constabulary role. Having no hand in creating the moral law, he is occupied in enforcing it. Logic prevails, or if not logic, then something in the laws of right and wrong that enforces their binding sense.

This is an attractive position, one that philosophers would wish to embrace, since it preserves some sense of a moral order without compromising their consensual position that their chief business is to decline an appeal to a supernatural order. And yet it is very difficult to find a way in which to justify the view that moral principles reflect some underlying cosmic necessity. They are no more like the laws of logic or mathematics than the laws of physics. Although some moral principles do appear universal in every human society, both in Nazi Germany and in Soviet Russia, societies were constructed in which familiar moral principles were inverted or discarded. To the extent that these societies survived, before they were destroyed by war or incompetence, they seemed perfectly able to flourish, their leaders never for a moment troubled by the thought that killing a great many people involved them in some form of intellectual inconsistency.

There remains nothing as a possibility in thought, if only by a process of elimination, and nothing is the preferred possibility in moral thought for the same reason it is the preferred possibility in physical thought: If logic is unavailing, then better nothing than God. This is just what Simon Blackburn means by refusing appeal to a supernatural order.

Nothing in moral philosophy has a familiar face. It is the position expounded both by freshmen in philosophy classes and all the enemies of humanity. We do not believe in any absolute moral truths, my students have always told me, although truths about grading seem a remarkably curious exception. Who could fail to hear the inner voice connecting this form of moral relativism to Himmler's? He, too, was a great believer in nothing, and nothing is just what so many scientific atheists believe as well.

What else is left?

❧

Like so many other positions, moral relativism has been promoted from the back of the college classroom to its podium. "The West," the philosopher Richard Rorty writes, "has cobbled together, in the course of the last two hundred years, a specifically secularist moral tradition—one that regards the free consensus of the citizens of a democratic society, rather than the Divine Will, as the source of moral imperatives." The words *the free consensus,* although sonorous, come to nothing more than the declaration that just so long as there is rough agreement within society, what its leaders say goes. This was certainly true of Nazi Germany. Many details of the final solution were kept hidden, but the view that the Jews of Europe were a problem requiring solution was so widespread in German society as to have appeared a commonplace. *Die Juden sind unser Unglück,* as a thick-fingered German butcher might have said—The Jews are our misfortune. The decision physically to kill them all expressed very nicely "the free consensus" of Germany's citizens. Had it not, the final solution could never have taken place. It did not reflect the consensus of citizens in Denmark, Italy, or Bulgaria, and in those countries there was no final solution, there was no mass deportation, and there were no extermination camps, and in all three cases, Nazi officials were left muttering in frustration at the fact that curiously enough these were places where people did not sufficiently appreciate the gravity of the Jewish problem.

Curiously enough.

Richard Rorty was to his great credit honest in facing the consequences of his own moral posture. He had no criticism to offer Nazi Germany beyond a personal sense of revulsion.

If moral imperatives are not commanded by God's will, and if they are not in some sense absolute, then what ought to be is a matter simply of what men and women decide should be. There is no other source of judgment.

What is this if not another way of saying that *if God does not exist, everything is permitted?*

These conclusions suggest quite justifiably that in failing to discover the source of value in the world at large, we must in the end retreat to a form of moral relativism, the philosophy of the fraternity house or the faculty dining room—similar environments, after all—whence the familiar declaration that just as there are no absolute truths, there are no moral absolutes.

Of these positions, no one believes the first, and no one is prepared to live with the second.

This is precisely the dilemma in which we find ourselves.

3

Horses Do Not Fly

DESPITE THE immense ideological power that it wields, the American scientific establishment has never trusted in its victory over organized religion (or anything else, for that matter). And for obvious reasons. On crucial matters of faith and morals, their margin of victory often seems paper-thin. Members of the National Academy of Sciences are by a large majority persuaded that there is no God, men and women in their millions that there is. *Thou, O king, sawest, and beheld a great image. This great image, whose brightness was excellent, stood before thee; and the form thereof was terrible. This image's head was of fine gold, his breast and his arms of silver, his belly and his thighs of brass, His legs of iron, his feet part of iron and part of clay.* Those who are religious contemplate this great image and see its head of gold; those who are not see its feet of clay. No division cuts deeper in the United States—or the world—or provokes a greater sense of mutual unease.

Looking thus toward those feet of clay, Sam Harris and Christopher Hitchens observe that many religious claims do not by the light of contemporary science appear to be true. Did Muhammad fly to Jerusalem on a horse named Borak? What an idea, Hitchens writes, observing alertly that "horses cannot and do not fly."

Addressing an audience of his Christian readers, Sam Harris asks them to consider the Moslem faith. He is quite certain that if they can find no reason to accept another man's beliefs, they will be moved at once to reject their own:

"Can you *prove* that Allah is not the one true God?"

"Can you *prove* that the Archangel Gabriel did not visit Muhammad in his cave?"

Richard Dawkins is less concerned to reject biblical miracles than to condemn the Deity for his hurtful insensitivity. "The God of the Old Testament," he writes, "is arguably the most unpleasant character in all of fiction: jealous and proud of it; a petty, unjust, unforgiving control-freak; a vindictive, blood thirsty ethnic cleanser; a misogynistic, homophobic, racist, infanticidal, genocidal, filicidal, pestilential, megalomaniacal, sadomasochistic, capriciously malevolent bully."

These are, to my way of thinking, striking points in God's *favor,* but opinions, I suppose, will vary.

It hardly matters. What is at issue is not so much the character of the Deity but his existence.

And the question *I* am asking is not whether he exists but whether science has shown that he does not.

THE EVIDENCE OF THINGS NOT SEEN

Faith, it is said in Hebrews 11.1, "is the substance of things hoped for, the evidence of things not seen." This is an interesting assertion, chaining as it does the concepts of faith, hope, evidence, and appearance. But in a sense, Hebrews 11.1 ratifies a triviality. We can make no sense either of daily life *or* the physical sciences in terms of things that are *seen*. The past has gone to the place where the past goes; the future has not arrived. We remember the one; we count on the other. If this is not faith, what, then, is it?

If religious belief places the human heart in the service of an unseen world, the serious sciences have since the great revolution of the seventeenth century done precisely the same thing. Mathematical physics has the narrative shape of a quest; physicists have placed their faith in the idea that deep down the universe is coordinated by a great plan, a rational system of organization, a hidden but accessible scheme, one that when finally seen in all its limpid but austere elegance, will flood the soul with gratitude. "All we [physicists] wish to do," Gerard 't Hooft has remarked, "is marvel at Nature's beauty and simplicity. We have seen and tasted the beauty, simplicity and universality of our latest theories. . . . We are now trying to uncover more of that. It is our belief that there is more." Our belief—meaning our *faith*.

Every scientist since Newton has placed his allegiance in the world beyond the world. In his remarkable treatise *The Road to Reality*, Roger Penrose quotes a letter from the mathematician Richard Thomas of the Imperial College in London. What is one to make, Penrose asks, of the remarkable, strange, and baffling mathematical results that have appeared in theoretical physics over the past twenty years or so? Thomas's reply is instructive and it is quite moving. "To a mathematician," he writes, "these things cannot be coincidence, they must come from a higher reason. *And that reason is the assumption that this big mathematical theory describes nature*" (italics added).

Western science is *above all* the substance of things hoped for, the evidence of things not seen.

Curiously enough, while Western science is saturated in faith, Western scientists remain incapable of seeing that faith itself, whether religious *or* scientific, is inherently vulnerable to doubt. Writing on his blog, the physicist Clifford Johnson observed that "failure is a possibility in any worth-while endeavor." True enough. It is. He went on to conclude that "this is an important distinction between scientific truth-searching and religious truth-searching where failure is not an option."

What a universe of careless contempt is expressed by these words. *Failure not an option?* And in the search for God? The world of sin and suffering is filled with those who have lost their religious faith, or given it up, or found the search impossible to sustain, or seen in pleasure a substitute for prayer, or as the hands of the clock crawled through the dark hours of the night, thought with a certain despair that it would be better not to search, and so not to doubt, and so not to be?

When Kierkegaard wrote about the sickness unto death, he was not remarking on a bronchial infection.

EVIDENCE

It is wrong, the nineteenth-century British mathematician W. K. Clifford affirmed, "always, everywhere, and for anyone, to believe anything upon insufficient evidence." I am guessing that Clifford believed what he wrote, but what evidence he had for his belief, he did not say.

Something like Clifford's injunction functions as the premise in a popular argument for the inexistence of God. If God exists, then his existence is a scientific claim, no different in kind from the claim that there is tungsten to be found in Bermuda. We cannot have one set of standards for tungsten and another for the Deity. If after scouring Bermuda for tungsten, we cannot find any of the stuff, then we give up on the claim. By parity of reasoning, if it is wrong to believe anything upon insufficient evidence, and *if* there is insufficient evidence for the existence of God, then it *must* be wrong to believe in his existence.

There remains the obvious question: By what standards might we determine that faith in science is reasonable, but that faith in God is not? It may well be that "religious faith," as the philosopher Robert Todd Carroll has written, "is contrary to the sum of evidence," but if religious faith *is* found wanting, it is reasonable to ask for a restatement of the rules by which "the sum of evidence" is computed. Like the Ten Commandments, they are difficult to obey but easy to forget. I have forgotten them already.

Perhaps this is because there are no such rules. The concept of sufficient evidence is infinitely elastic. It depends on context. Taste plays a role, and so does intuition, intellectual sensibility, a kind of feel for the shape of the subject, a desire to be provocative, a sense of responsibility, caution, experience, and much besides. Evidence in the court of public opinion is not evidence in a court of law. A community of Cistercian monks padding peacefully from their garden plots to their chapel would count as evidence matters that no physicist should care to judge. What a physicist counts as evidence is not what a mathematician generally accepts. Evidence in engineering has little to do with evidence in art, and while everyone can agree that it is wrong to go off half-baked, half-cocked, or half-right, what counts as being baked, cocked, or right is simply too variable to suggest a plausible general principle.

When a general principle *is* advanced, it collapses quickly into absurdity. Thus Sam Harris argues that "to believe that God exists is to believe that *I* stand in some relation to his existence *such that his existence is itself the reason for my belief*" (italics added). This sounds very much as if belief in God

could only be justified if God were to call attention conspic-
uously to Himself, say by a dramatic waggling of the divine
fingers.

If this is so, then by parity of reasoning again, one might
argue that to believe that neutrinos have mass is to believe
that I stand in some relationship to their mass such that their
mass is itself the reason for my belief.

Just how are those neutrinos waggling *their* fingers?

A neutrino *by itself* cannot function as a reason for my belief. It is a subatomic particle, for heaven's sake. What I believe
is a proposition, and so an abstract entity—*that* neutrinos have
mass. How could a subatomic particle enter into a relationship
with the object of my belief? But neither can a neutrino be the
cause of my belief. I have, after all, never seen a neutrino: not
one of them has ever gotten *me* to believe in *it*. The neutrino,
together with almost everything else, lies at the end of an immense inferential trail, a complicated set of judgments.

Believing as I do that neutrinos have mass—it is one of my
oldest and most deeply held convictions—I believe what I do
on the basis of the fundamental laws of physics and a con-
geries of computational schemes, algorithms, specialized pro-
gramming languages, techniques for numerical integration,
huge canned programs, computer graphics, interpolation methods, nifty shortcuts, and the best efforts by mathematicians
and physicists to convert the data of various experiments into
coherent patterns, artfully revealing symmetries and continuous narratives. The neutrino has nothing to do with it.

Within mathematical physics, the theory determines the evidence, and not the other way around. What sense could one make of the claim that top quarks exist in the absence of the Standard Model of particle physics? A thirteenth-century cleric unaccountably persuaded of their existence and babbling rapturously of quark confinement would have faced then the question that all religious believers now face: Show me the evidence. Lacking access to the very considerable apparatus needed to test theories in particle physics, it is a demand he could not have met.

In the face of experience, W. K. Clifford's asseveration must be seen for what it is: a moral principle covering only the most artificial of cases.

The existence of God is not one of them.

NATURALISM

Neither the premises nor the conclusions of any scientific theory mention the existence of God. I have checked this carefully. The theories are *by themselves* unrevealing. If science is to champion atheism, the requisite demonstration must appeal to something in the sciences that is not quite a matter of what they say, what they imply, or what they reveal.

In many respects the word *naturalism* comes closest to conveying what scientists regard as the spirit of science, the source of its superiority to religious thought. It is commended as an attitude, a general metaphysical position, a universal doctrine—and often all three. Rather like old-fashioned Swedish sunshine-and-seascape nudist documentaries, naturalism is a term that conveys an agreeable suggestion of healthful inevitability. What, after all, could be more natural than being natural? Carl Sagan's buoyant affirmation that "the universe is everything that is, or was, or will be" is widely understood to have captured the spirit of naturalism, but since the denial of this sentence is a contradiction, the merits of the concept so defined are not immediately obvious. Just who is arguing from the pulpit that everything is *not* everything?

A triviality having been affirmed, what follows surges into the badlands in which assertions remain unsupported by arguments. "Everything," the philosopher Alexander Byrne has remarked, "is a natural phenomenon." Quite so. But each of those natural phenomena is, Byrne believes, simply "an aspect of the universe revealed by the natural sciences." If what is natural has been defined in terms of what the natural sciences reveal, no progress in thought has been recorded. If not, what reason is there to conclude that everything is an "aspect of the universe revealed by the natural sciences"?

There is no reason at all.

<6?

If *naturalism* is a term largely empty of meaning, there is always *methodological* naturalism. Although naturalism is natural, methodological naturalism is even more natural and is, for that reason, a concept of superior grandeur. Hector Avalos is a professor of religious studies at Iowa State University, and an avowed atheist. He is a member in good standing of the worldwide fraternity of academics who are professionally occupied in sniffing the underwear of their colleagues for signs of ideological deviance. Much occupied in denouncing theories of intelligent design, he has enjoyed zestfully persecuting its advocates. "Methodological *naturalism*," the odious Avalos has written, "the view that *natural* phenomena can be explained without reference to super*natural* beings or events, is the foundation of the *natural* sciences."

Now a view said to be foundational can hardly be said to be methodological, and if naturalism is the foundation of the natural sciences, then it must be counted a remarkable oddity of thought that neither the word nor the idea that it expresses can be found in any of the great physical theories. Quite the contrary. Isaac Newton in writing the *Principia Mathematica* seemed curiously concerned to place rational mechanics on a foundation that has nothing to do with methodological naturalism. "The most beautiful system of the sun, planet and comets," he wrote, "could only proceed from the counsel and domination of an intelligent and powerful Being."

There is finally the claim that the universe is a closed causal system, the triplet of its three vaguely technical terms suggesting something more substantial by way of a definition. But to say that the universe is a causal system is hardly an improvement on the thesis that effects have causes, and if the universe is everything that there *is,* then to say that it is closed is only to observe that there is nothing beyond everything.

This is not a thesis calculated to set the blood racing.

MATTER

There is nothing in nature, ancient Greek atomists said, but atoms and the void, and while this claim has over the centuries been refined, it remains deep down the same. The end of the matter is matter. Materialists have always hoped that by going downward, they would at last reach the ultimate level of analysis and so the place where Nature reveals her ontological essentials by means of a finite number of elementary particles. This is a matter of faith. It is entirely possible that there may be as many elementary particles as there is funding available to investigate them.

The advantage of materialism as a doctrine is that it sanctions an easy argument for atheism. Either the Deity is a material object or he is not. If he is, then he is just one of those things, and if he is not, then materialism could not be true. But if God is just one of those things, what is his interest? And if materialism is false, why are we arguing?

Whatever the merits of this argument, the world of matter revealed by the physical sciences does not serve to endow materialism with a familiar face. The universe in its largest aspect is the expression of curved space and time. Four fundamental forces hold sway. There are black holes and various infernal singularities. Popping out of quantum fields, the elementary particles appear as bosons or fermions. The fermions are divided into quarks and leptons. Quarks come in six varieties, but they are never seen, confined as they are within hadrons by a force that perversely grows weaker at short distances and stronger at distances that are long. There are six leptons in four varieties. Depending on just how things are counted, matter has as its fundamental constituents twenty-four elementary particles, together with a great many fields, symmetries, strange geometrical spaces, and forces that are disconnected at one level of energy and fused at another, together with at least a dozen different forms of energy, all of them active.

This is not an ontology that puts one in mind of a longshoreman's view of the material world. It is remarkably baroque. And it is promiscuously catholic. For the atheist persuaded that materialism offers him a no-nonsense doctrinal affiliation, materialism in this sense comes to the declaration of a barroom drinker that he will have whatever *he's* having, no matter *who* he is or *what* he is having. What *he* is having is what he always takes, and that is any concept, mathematical structure, or vagrant idea needed to get on with it. If tomorrow, physicists determine that particle physics requires access to the ubiquity of the body of Christ, that doctrine would at once be declared a physical principle and treated accordingly.

THE SCIENTIFIC METHOD

The scientific method has acquired a certain hold on the popular imagination. Every adult remembers something about the scientific method from high school classes; it figures prominently in textbooks with such titles as *Reasoning Together,* and it is a polemical bruiser in its weight class, useful under circumstances when members of the scientific community are persuaded they are under attack. It is then that the determination is made that members of the public have failed to understand the scientific method or properly to revere it. No effort need be made actually to exhibit the method or tie it to an argument.

All of this provides a richly satisfying spectacle.

Here is one account, an Internet staple. To apply the scientific method

1. Observe some aspect of the universe.
2. Form a hypothesis that potentially explains what you have observed.
3. Make testable predictions from that hypothesis.
4. Make observations or experiments that can test those predictions.
5. Modify your hypothesis until it is in accord with all observations and predictions.

Not a single one of these five sentences makes the slightest sense, but rather than go through the list, let me observe only that it is portable in its power, and applies pretty much to any human undertaking.

"Through extensive observation, I found a Common Denominator among all Golfers, and once I finally realized what that Common Denominator was, I just couldn't believe how 'obvious' and simple it all was. As in any puzzle or 'discovery,' the idea was right in front of my eyes all the time!

"The Common Denominator I discovered was that *all* golfers who break 80 regularly are good, or at least fairly good at a certain Element in the golf swing, and *all* golfers who don't break 80 are bad at that same thing. From this simple observation came the obvious conclusion that this Element was the first and most important thing that needed to 'be in' and to be learned in order to shoot in the 70s!

"This method is based upon this observable fact (Common Denominator). So the next thing to do was to test this idea to see if this method really worked. And the answer? Yes, it did, and in Spades! I saw changes in minutes and hours, and huge big smiles on people's faces. Handicaps were being cut in half within weeks!"

I will draw down the current of charity over this scene. Golf has no method beyond the trivial.

Neither does science.

NOTHING BUT THE TRUTH

What remains of the ideology of the sciences? It is the thesis that the sciences are true—who would doubt it?—and that *only* the sciences are true. The philosopher Michael Devitt thus argues that "there is only one way of knowing, the empirical way that is the basis of science." An argument against religious belief follows at once on the assumptions that theology is not science and belief is not knowledge. If by means of this argument it *also* follows that neither mathematics, the law, nor the greater part of ordinary human discourse have a claim on our epistemological allegiance, they must be accepted as casualties of war.

Declarations of this sort have been common in the history of philosophy since the eighteenth century. In *An Enquiry Concerning Human Understanding*, David Hume argued that "if we take in our hand any volume; of divinity or school metaphysics, for instance; let us ask, Does it contain any abstract reasoning concerning quantity or number? No. Does it contain any experimental reasoning concerning matter of fact and existence? No. Commit it then to the flames: For it can contain nothing but sophistry and illusion!" Analytical philosophers have been eager to commit books to the flames ever since, rather an odd vocational choice, all things considered. Whatever the vigor with which Hume advanced his views, arguments such as his when self-applied self-destruct. Hume's remarks, after all, contain neither "abstract reasoning concerning quantity or number" nor "experimental reasoning concerning matters of fact and existence." They are what they seem, and that is at once arrogant and uninteresting.

The attempt to find an argument powerful enough to

paralyze distasteful doctrines, while remaining insusceptible to its own effects, has continued into our time. In his well-known essay "Two Dogmas of Empiricism," W. V. O. Quine argued that the distinction between science and philosophy was an illusion. Philosophers were pleased since Quine appeared to offer the access to a form of prestige that previously they had been denied. If there is no distinction between science and philosophy, they reasoned, then we must be scientists. That they might by the same logic be *nothing* was an alternative that did not receive wide favor. Physicists, on the other hand, seemed remarkably unenthusiastic about welcoming philosophers as fellow scientists. "The philosphers," Richard Feynman observed, "are always on the outside making stupid remarks." Critics observed—correctly—that Quine's argument seemed to affirm what it was most concerned to deny. In arguing that there was no distinction between science and philosophy, Quine was arguing as a philosopher, and he was making a philosophical argument. If this is science, anything is. And if it is not, so much the worse for the philosophers, who once again would appear to be "on the outside making stupid remarks."

An ideological system whose proponents are persuaded that access to the truth is in their hands requires an equally general defense against criticism. As one might expect, it lies close at hand. The sciences, many scientists argue, require no criticism because the sciences comprise a uniquely *self*-critical institution, with questionable theories passing constantly before stern appellate review. Judgment is unrelenting. And impartial. Individual scientists may make mistakes, but like the Communist Party under Lenin, science is infallible because its judgments are collective. Critics are unneeded, and since they are unneeded, they are not welcome.

A system so conceived always works to the satisfaction of those who have conceived it. In *Six Impossible Things Before Breakfast,* the biologist Lewis Wolpert, who is resolutely prepared to dismiss religious thought as superstition, writes that "scientific beliefs are special, and different from any other kind of thinking," inasmuch as scientific beliefs "are not programmed into our brains." To say that scientific beliefs are special is to suggest, of course, that only specialists may assess them. To say that religious beliefs are programmed into our brains is to say that like the gag reflex, they cannot be controlled. But if scientific beliefs are not programmed into our brains, why assume that religious beliefs are? And if they are not, why assume that "scientific beliefs are special"?

These questions are rhetorical. No one is disposed to ask them within the scientific community, and the scientific community is not disposed to acknowledge answers to questions it is not disposed to ask.

⟆

The idea that we must turn to the sciences in order to assess our religious beliefs owes much to the popular conviction that so long as we are turning, where else are we to turn *to*? The proper response is a question in turn. Why turn at all? And if we must turn, why turn in the wrong direction? To ask of the physical sciences that they assess the Incarnation, or any other principle of religious belief, is rather like asking of a powerful Grand Prix racing car that it prove itself satisfactory in doing service as a New York taxicab.

The claim that the existence of God should be treated as a scientific question stands on a destructive dilemma: If by science one means the great theories of mathematical physics, then the demand is unreasonable. We cannot treat *any* claim in this way. There is no other intellectual activity in which theory and evidence have reached this stage of development.

If, on the other hand, the demand means merely that one should treat the existence of God as the existence of anything would be treated, then we must accept the fact that in life as it is lived beyond mathematical physics, the evidence is fragmentary, lost, partial, and inconclusive. We do what we can. We grope. We see glimmers.

At times, the light. "The very instant I heard my father's cry calling unto me, my heart bounded in recognition."

At times, the darkness. "A blank was there instead of *it*. . . . Life had become curiously dead and indifferent."

And as is always the case, someone may be found honest enough to blurt out the truth.

Is there a God who has among other things created the universe? "It is not by its conclusions," C. F. von Weizsäcker has written in *The Relevance of Science*, "*but by its methodological starting point* that modern science excludes direct creation. *Our methodology would not be honest if this fact were denied* ... such is the *faith* in the science of our time, and which we all share" (italics added).

In science, as in so many other areas of life, faith is its own reward.

4

The Cause

T HE COSMOLOGICAL argument emerges from a simple question and its answer.

The question:

What caused the universe?

The answer:

Something.

Some form of this argument has appeared in every human culture. It is universal. For all men, this argument sometimes appears sound, and for some men, always. Is this a surprise? We are talking, after all, about the existence of God, and if the issue were easily decided, we would not be talking. The

medieval Arabic argument known as the kalam is an example
of the genre.

Its first premise:
> *Everything that begins to exist has a cause.*

And its second:
> *The universe began to exist.*

And its conclusion:
> *So the universe had a cause.*

This is not by itself an argument for the existence of God.
It is suggestive without being conclusive. Even so, it is an
argument that in a rush covers a good deal of ground carelessly denied by atheists. It is one thing to deny that there is a
God; it is quite another to deny that the universe has a cause.
What remains, if the universe does have a cause, is the gap
between *what* brought the universe into existence and traditional conceptions of the deity. This is no trivial matter. Nonetheless, the cosmological argument succeeds in displacing the
burden of proof from its starting point *(Is there a God?)* to a
place much later in the argument *(Is it right and proper to think
that the cause of the universe is God?)*.

THOMAS AQUINAS

The most powerful statement of the cosmological argument is due to Thomas Aquinas, the largest intellectual
personality of the thirteenth century. A master of the high

scholastic method—Latin, liturgy, and logic—Aquinas synthe-
sized Aristotelian philosophy and the doctrines of the Catholic
Church so successfully that to this day, the style of argument
adopted by the Vatican represents his influence. Nonetheless,
Aquinas is not an easy philosopher to read, and he is not fashionable. This is not a decisive point in his favor, but it is difficult to ignore.

Aquinas was born in 1225 in southern Italy and died fifty
years later in a Cistercian monastery in northern Italy. His life
coincided with a period of great brilliance in European art,
architecture, law, poetry, philosophy, and theology. Commentators who today talk of the *dark* ages, when faith instead of
reason was said ruthlessly to rule, have for their animadver-
sions only the excuse of perfect ignorance.

Both Aquinas's intellectual gifts and his religious nature
were of a kind that is no longer commonly seen in the Western
world. Devoted and obedient, he approached the mansion of
the Catholic faith with the confidence of someone sure of his
welcome at the door and of his comfort within its rooms. The
natural world did not attract his attention. He was not curious.
He neither conducted experiments nor imagined that it would
be worthwhile to do so. His genius was organizational and logical and even, in its largest aspect, legal. His masterpiece and
his monument, the *Summa Theologica*, contains 38 treatises, and
deals with 612 separate questions, subdivided into 3,120 separate sections. In all, the work asks and answers ten thousand
questions. It is a cathedral in thought, inviting admiration but

not affection. Those who reject atheism still find it difficult to accept Aquinas. He is in his sensibility now alien.

On December 6, 1273, Aquinas, while attending mass, fell into a prolonged and rapturous mystical state. Thereafter, he ceased to write. When urged by officials of the Catholic Church to continue his work on the *Summa*, which he had left unfinished, he replied, "I can do no more. Such secrets have been revealed to me that all I have written now appears to be of little value."

ᔟ

Aquinas addresses the cosmological argument in Article 3 of Question 2 of the first part of the *Summa*. Question 2 is called "The Existence of God," and Article 3 asks the question whether God exists. Aquinas begins by offering a powerful and lucid defense of atheism.

"It is superfluous to suppose," Aquinas argues, "that what can be accounted for by a few principles has been produced by many." This constraint is now familiar as Occam's Razor, even though William of Occam lived and wrote after Aquinas's death.

"But it *seems*," Aquinas at once adds, "that everything we see in the world can be accounted for by other principles, supposing that God did not exist."

Other principles?

Just so. "All natural things can be reduced to one principle, which is nature, and all voluntary things can be reduced to one principle, which is human reason, or will."

It follows, Aquinas concludes provisionally, that "[t]here is no reason to suppose God's existence."

This is a conclusion that Aquinas is prepared to reject with all the force of his faith and genius. The existence of God may be demonstrated; it is subject to proof, and if not proof, then to argument. It follows that not everything in nature *can* be accounted for by "other principles."

The economies of thought offered by Occam's Razor are an illusion.

᠂᠊ᢀ

We understand things in nature, Aquinas observes, by grasping as best we can causes and their effects: the match that lights the fire, the chill that sets one's teeth to chattering, the water that slakes thirst. "In the world of sense," as Aquinas says, "there is an order of efficient causes." But just as no man can be his own father, no effect can be its own cause. A series of effects preceded by their causes forms a luminous metaphysical trail going backward into the past, because, as Aquinas argues, causes must *precede* their effects.

Can a series of this sort be *infinitely* continued, so that it simply disappears into the loom of time?

Aquinas argues that when it comes to causes, "it is not possible to go on to infinity, because in all . . . causes following in order, the first is the cause of the intermediate cause, and the intermediate is the cause of the ultimate cause."

If a series of causes does not start, it cannot get going, and if it does not get going, then there will be no intermediate causes, and if there are no intermediate causes, then over here, where we have just noticed that a blow has caused a bruise, there is no explanation for what is before our eyes. Either there is a first cause or there is no cause at all, and since there are causes at work in nature, there must be a first. The first cause, Aquinas identified with God, because in at least one respect, a first cause exhibits an important property of the divine: *It is uncaused.*

This is a weak but not an absurd argument, and while Aquinas's conclusion may not be true, objections to his argument are frequently inept. Thus Richard Dawkins writes that Aquinas "makes the entirely unwarranted *assumption* that God is immune to the regress." It is a commonly made criticism. Lumbering dutifully in Dawkins's turbulent wake, Victor Stenger makes it as well. But Aquinas makes no such assumption, and thus none that could be unwarranted. It is the *conclusion* of his argument that causes in nature cannot form an infinite series.

A far better objection has long been common in the philosophical literature: While an infinite series of causes has no first cause, it does not follow (does it?) that any specified effect is without a cause. Never mind the first cause. This blow has caused that bruise. The chain of causes starting with the blow may be chased into the past to any finite extent, but no matter how far back it is chased, effects will always have causes. Why, then, is that first cause so very important?

But this is a counterargument at which common sense is inclined to scruple. Seeing an endless row of dominoes toppling before our eyes, would we without pause say that no first domino set the other dominoes to toppling?

Really?

The give-and-take of these arguments is worthy of respect, but it no longer compels attention. In the eight hundred years following the publication of the *Summa,* the philosophers have had their say, but they have been overtaken by events. The argument that Aquinas wished to make on metaphysical grounds has been made in other terms and in other ways, and in particular a form of the cosmological argument has appeared in the very place one might least expect it to appear: contemporary physical cosmology.

THE THRESHOLD OF THEOLOGY

The universe, orthodox cosmologists believe, came into existence as the expression of an explosion—what is now called the Big Bang. The word *explosion* is a sign that words have failed us, as they so often do, for it suggests a humanly comprehensible event—a *gigantic* explosion or a *stupendous* eruption. This is absurd. The Big Bang was not an event taking place at a time or in a place. Space and time were themselves created by the Big Bang, the measure along with the measured.

If the image of an ordinary explosion is inadequate to the Big Bang, the words themselves—*the Big Bang*—have by themselves a disturbing power. They suggest the most ancient of human intuitions, and that is the connection between sexual and cosmic energies. The words may have been chosen whimsically; they were not chosen accidentally.

Whatever its name, as far as most physicists are concerned, the Big Bang is now a part of the established structure of modern physics. From time to time, it is true, the astrophysical journals report the failure of observation to confirm the grand design. It hardly matters. The physicists have not only persuaded themselves of the merits of Big Bang cosmology, they have persuaded everyone else as well. The Big Bang has come to signify virtually a universal creed, men and women who know nothing of cosmology convinced that the rumble of creation lies within reach of their collective memory.

❧

If the Big Bang expresses a new idea in physics, it suggests an old idea in thought: *In the beginning God created the heaven and the earth.* This unwelcome juxtaposition of physical and biblical ideas persuaded the astrophysicist Fred Hoyle, an ardent atheist, to dismiss the Big Bang after he had named it. In this he was not alone. Many physicists have found the idea that the universe had a beginning alarming. "So long as the universe had a beginning," Stephen Hawking has written, "we could suppose it had a creator." *God forbid!* Nonetheless, there *is* a very natural connection between the fact that the universe had a beginning and the hypothesis that it had a creator. It is a connection so plain that, glowing with its own energy, it may be seen in the dark. Although questions may be raised about what it means, the connection itself cannot be ignored. "The best data we have concerning the big bang," the Nobel laureate Arno Penzias remarked, "are exactly what I would have predicted, had I nothing to go on but the five books of Moses, the Psalms, the Bible as a whole."

Remarks such as this traveled far afield. They were repeated gratefully by men and women persuaded that at last cosmology had made some sort of sense. They appeared in the *New York Times.* Physicists quickly came to their senses. They discovered elaborate reasons to avoid the obvious, not least of which, the fact that the obvious *was* obvious. For more than a century, physicists had taken a manful pride in the fact that theirs was a discipline that celebrated the weird, the bizarre, the unexpected, the mind-bending, and the recondite. Here was a connection that any intellectual primitive could at once grasp: The universe had a beginning, thus something must have caused it to begin. Where would physics be, physicists asked themselves, if we had paid the slightest attention to the obvious?

In this, the physicists were immeasurably assisted by the philosophers, their traditional enemies, of course, who aided in the work at hand by writing very elegant papers demonstrating that if the universe had a beginning, it was not a beginning that really began. The philosopher Adolf Grünbaum of the University of Pittsburgh was a master of this approach. If the universe did not have a beginning, his papers did not have an end. Fair is fair. Physicists who had been struggling to make precisely the same point welcomed such philosophical efforts with the relief a stutterer might show on having his interlocutor blurt out the stammered word.

All might have been well, or at least better than it turned out to be, had the Big Bang been another one of those tedious ideas that flicker luridly for a moment and then wink out. There are so many of them. But quite the contrary proved to be the case. Over the course of more than half a century—a very long time in the history of the physical sciences—inferences gathered strength separately, and when combined they gathered strength in virtue of their combination.

One line of inference was observational; the second, theoretical; the two together, irresistible.

⚜

The observations that made the hypothesis of the Big Bang plausible were derived from a study of the heavens. They had some of the brute power of something *seen*. This is an exaggeration, of course. Observations themselves depend on a network of theoretical assumptions, but in the case of *these* observations, their theoretical structure belonged to a part of physics that was well understood. No astronomer scanning the cosmos, for example, doubted much the plainest of plain facts about light. No matter how it appears, light represents an undulation of the electromagnetic field. Its source is the excitable atom itself, with electrons bouncing from one orbit to another and releasing energy as a result. If this is so, it follows that each atom has a spectral signature, a distinctive electromagnetic frequency. The light that streams in from space thus must reveal something about the composition of the galaxies from which it was sent. In the early years of the twentieth century, the characteristic signature of hydrogen was detected in various far-off galaxies. Examining a very small sample of twenty or so galaxies, the American astronomer Vesto Slipher observed that the frequency of their hydrogen atoms was *shifted* to the red portion of the spectrum. Using a far more sophisticated telescope than any at Slipher's disposal, Edwin Hubble made the same discovery in the early 1930s, and unlike poor Slipher, *he* knew he had struck gold.

The galactic redshift, Hubble realized, was an exceptionally vivid cosmic clue, a bit of evidence from far away and long ago, and as with all clues its value lay in the questions it prompted. Why should galactic light be shifted to the red and not the blue portions of the spectrum? Why, for that matter, should it be shifted at all?

These very simple questions received an equally simple answer, with neither question nor answer ever transgressing the margins of plain physical sense. The pitch of a siren is altered as a police car disappears down the street, the sound

waves carrying the noise stretched by the speed of the car itself. This is the familiar Doppler effect. Something similar
explains the redshift of the galaxies. Distortions in their spectral signature arise because these monsters of the night are
receding into the depths. But a universe whose galaxies are
receding is one that is expanding.

The inference to the Big Bang now follows. A universe
that is expanding is a universe with a clear path into the past.
If things are now far apart, they must at one point have been
close together; and if things were once close together, they
must at one point have been *hotter* than they are now, the contraction of space acting to compress its constituents like a
vise, and so increase their energy. The retreat into the past
ends in a state in which material particles are at *no* distance
from one another and the temperature, density, and curvature
of the universe are infinite. Such a state is known as a singularity, and in the case of the cosmos it is known as the Big
Bang singularity.

The cone tapering into the past must end. The lines of
sight converge. The universe had a beginning.

❦

When the facts about an expanding universe became known,
physicists at once realized that they had become known in the
right place and at the right time. In 1915, Albert Einstein had
published his theory of general relativity. The theory represented the culmination of a revolution in physical thought

that he had initiated in 1905 with the publication of his theory of special relativity. The general theory of relativity encompassed Einstein's account of gravitation, but because gravitation extends throughout space and time as a universal force, his theory was simultaneously a kind of cosmic blueprint, a way of grasping by mathematical means the ultimate structure of the cosmos.

General relativity forges a far-flung connection between the geometry of space and time and the presence of matter. Events within Einstein's majestic theory are designated by four numbers. Three of these numbers indicate where the event is, and the fourth measures when it is there. Physicists very much enjoy suggesting that the world of four dimensions is so inaccessible that entrance is generally denied the mathematically uninitiated. But although those four dimensions are important, the underlying concept is simple. After all, we locate an event in terms of both *where* it took place and *when* it took place. Where was JFK assassinated? Three numbers provide the answer (longitude, latitude, and height). And when? One number is sufficient. To have grasped this much is to have grasped everything. (And as long as secrets are being imparted, when physicists talk of ten-dimensional or eleven-dimensional spaces, nothing deeper is at issue.) To a fused form of space and time in which points are identified by four numbers, Einstein assigned a *variable* geometrical structure. Squeezing a solid rubber ball produces the same effect, although not, of course, on the same scale. *Now* it is perfectly round. A squeeze is administered, and *then* it becomes deformed. With the re-lease that follows, its shape changes again, and with its shape, its geometry.

If the geometry of space and time *is* variable, Einstein conjectured, this must affect the way in which material objects move through the medium *of* space and time. A beam of light crossing an otherwise empty universe travels in a straight line. If a massive star is placed in its path, the light beam will curve, almost as if its graceful swerve were intended to avoid a collision. By the same token, an observer falling toward the earth, his failed parachute dangling uselessly behind him and an ignominious plop forthcoming, is doing nothing more than traveling along his natural path through space and time. He appears to be accelerating because the earth has distorted the geometry through which he is falling.

Einstein's theory of general relativity involves the cohabi-tation of these conceptual partners. Material objects influence space and time by deforming their geometry. Space and time influence material objects by changing their path. The rela-tionship goes both ways.

The field equation that Einstein introduced in 1915 is a majestic identity in which curvature, on the one side, and mass, on the other, are placed in the balance and found equal. Ein-stein had hoped that the equations of general relativity would determine a single world model, and like virtually every other physicist, he believed that his cosmic blueprint would reveal a universe that had neither a beginning nor an end. Searching for what he wished to find, Einstein discovered a solution to his own equations that specified just such a universe, the great thing having been there from the infinite past and destined to be there into the infinite future. For reasons that he could never make clear, Einstein found a universe so conceived particularly satisfying. Friends of his who knew him well have suggested (to me) that to the end of his life, Einstein regarded an expanding universe with a certain fastidious distaste.

Einstein had hoped that the equations of his great theory would specify only one cosmic blueprint. In this he was destined to be disappointed. Months after he discovered one solution of the field equations, Willem de Sitter discovered another. In de Sitter's universe, there is no matter whatsoever, the place looking rather like a dance hall in which the music can be heard but no dancers seen. Dismissed at the time, the de Sitter universe has recently enjoyed a revival in quantum cosmology. It is easy to describe, easy to find, and like the diligent Dutch themselves, endlessly useful.

In the 1920s, both Aleksandr Friedmann and Georges Lemaître discovered the solutions to the field equations that have dominated cosmology ever since, their work coming to amalgamate itself into a single denomination as Friedmann-Lemaître (FL) cosmology. To Einstein's pained surprise, FL cosmology indicated that the universe was either expanding or contracting, a conclusion nicely in accord with Hubble's observation but profoundly in conflict with models of the universe in which the universe remained resolutely unchanging.

Having been joined at the fulcrum of observation and theory, Big Bang cosmology has been confirmed by additional evidence, some of it astonishing. In 1963, the physicists Arno Penzias and Robert Wilson observed what seemed to be the living remnants of the Big Bang—and after 14 billion years!— when in 1962 they detected, by means of a hum in their equipment, a signal in the night sky they could only explain as the remnants of the microwave radiation background left over from the Big Bang itself.

More than anything else, this observation, and the inference it provoked, persuaded physicists that the structure of Big Bang cosmology was anchored into fact.

The wheel had come full circle.

THE INESCAPABLE BEGINNING

If both theory and evidence suggested that the universe had a beginning, it was natural for physicists to imagine that by tweaking the evidence and adjusting the theory, they could get rid of what they did not want. Perhaps the true and the good universe—the one without a beginning—might be reached by skirting the Big Bang singularity, or bouncing off it in some way? But in the mid-1960s, Roger Penrose and Stephen Hawking demonstrated that insofar as the backward contraction of the universe was controlled by the equations of general relativity, almost all lines of conveyance came to an end.

The singularity was inescapable.

This conclusion encouraged the theologians but did little to ease physicists in their own minds, for while it strengthened the unwholesome conclusions that Big Bang cosmology had already established, it left a good deal else in a fog. In many ways, this was the worst of all possible worlds. Religious believers had emerged from their seminars well satisfied with what they could understand; the physicists themselves could understand nothing very well.

The fog that attended the Penrose-Hawking singularity theorems (there is more than one) arose spontaneously whenever physicists tried to determine just what the singularity signified. At the singularity itself, a great many physical parameters zoom to infinity. Just what is one to make of infinite temperature? Or particles that are no distance from one another. The idea of a singularity, as the astronomer Joseph Silk observed, is "completely unacceptable as a physical description of the universe. . . . An infinitely dense universe [is] where the laws of physics, and even space and time, break down."

Does the singularity describe a physical state of affairs or not?

Tell us.

If it does, the description is uninformative by virtue of being "completely unacceptable." If it does not, the description is uninformative by virtue of being completely irrelevant. But if the description is either unacceptable or irrelevant, what reason is there to believe that the universe began in an initial singularity? Absent an initial singularity, what reason is there to believe that the universe *began*?

If the universe did not begin, but had nonetheless only a finite temporal extent, what on earth are we to think at all?

It may seem that a conclusion has been reached that will appeal to physicists and religious believers alike: *Nothing can be said.* Those who believe in God and those who do not may resolve their differences by agreeing to say nothing. There is nonetheless a striking point at which Big Bang cosmology and traditional theological claims intersect. The universe has *not* proceeded from the everlasting to the everlasting. The cosmological beginning may be obscure, but the universe is finite in time. This is something that until the twentieth century was not known. When it became known, it astonished the community of physicists—and everyone else. If nothing else, the facts of Big Bang cosmology indicate that one objection to the argument that Thomas Aquinas offered is empirically unfounded: Causes in nature do come to an end. If science has shown that God does not exist, it has not been by appealing to Big Bang cosmology. The hypothesis of God's existence and the facts of contemporary cosmology are *consistent*.

The uncertainties surrounding the origin of the universe have led certain writers to find comfort in a companionship with Aquinas they would not otherwise dream of enjoying. In writing about the first cause to which Aquinas appealed, and which he identified with God, Richard Dawkins argues that "it is more parsimonious to conjure up, say, a 'Big Bang singularity,' or some other physical concept as yet unknown" to account for the existence of the universe. The word *parsimonious* is meaningless in context: Whatever it might denote, how could it be measured? But *conjure* is the right verb, suggesting as it does both misdirection and inattention. Misdirection: The Big Bang singularity does not represent a *physical* concept, because it cannot be accommodated by a physical theory. It is a point at which physical theories give way. Inattention: The concept in which Dawkins *has* placed his confidence is something that is either infinite and inscrutable, or otherwise unknown. Men have come to faith on the basis of far less. This is, I suppose, not surprising. His atheism notwithstanding, Dawkins believes that he is a "deeply religious man." He simply prefers an alien cult.

"Perhaps the best argument in favor of the thesis that the Big Bang supports theism," the astrophysicist Christopher Isham has observed, "is the obvious unease with which it is greeted by some atheist physicists. At times this has led to scientific ideas, such as continuous creation or an oscillating universe, being advanced with a tenacity which so exceeds their intrinsic worth that one can only suspect the operation of psychological forces lying very much deeper than the usual academic desire of a theorist to support his or her theory."

5

The Reason

I am that I am. —EXODUS 3:14

THE COSMOLOGICAL argument just given covers familiar ground: God is a cause. But God enters the troubled human imagination in a second way, and that is as the answer to the question *why* the universe exists at all. Something deeper is at issue, and so something deeper is wanted. Even if we understood how the universe came into existence, the question why it exists and why it continues to exist would remain.

At some moment in the unrecoverable past, the battle-ready Hebrews understood that the scattered deities of the Near Eastern world were manifestations of a single God. "Hear, O Israel: The Lord our God, the Lord is one!"

If God is one, he is one *absolutely*, the Hebrew Bible affirms, because not only does he exist, he *must* exist. The five simple words of the declaration in Exodus—"I am that I am"—suggest that God's existence is necessary. Being what He is, God could not fail to be who He is, and being who He is, God could not fail to *be*.

This is the heart of a second cosmological argument. It draws a connection between the existence of the universe and the existence of the Deity. The argument is not simple, and it is by no means conclusive.

᪥

Everything that exists has a precarious hold on being. Here today, gone tomorrow is more than an adage; it is a principle of metaphysics. We have an uncommon ability mentally to shuffle things in and out of existence; but applied so easily to others, this power cannot be self-applied. No matter with what determination we stare into the void, the staring itself makes the effort an exercise in irrelevance. *Who* is staring? If we cannot imagine a world without us (and so in my case a world gone mad with grief), we can give our reluctant assent to the proposition that things might continue in our absence.

Aquinas applies this argument to the universe, because he can see no reason to suppose that its existence is guaranteed. If it might not exist, why, then, *does* it exist?

Why indeed?

There now follows a remarkable, a bold, but a problematic

step in the argument: If it is possible that something might not exist, Aquinas asserts, then it is certain that at some time it did not exist. In this, Aquinas was reprising a view of possibility that may be traced back to the Greek philosopher Diodorus.

But if the universe did not exist at some moment of time, then it emerged from absolutely nothing. The universe is everything that there is. What beyond nothing is left to explain its promotion from inexistence to existence?

This, Aquinas observes, is incoherent. *Ex nihilo nihil fit.* From nothing, nothing, as ancient writers said. Because it is impossible to understand the emergence of something from nothing, Aquinas concludes, something must have acted to bring the universe into existence. That something, the argument continues, could have been contingent or necessary. If contingent, we are no further advanced. We have simply chased perplexities into the past. If not contingent, then necessary. When it comes to things that exist necessarily, it is wasteful to assume more than one. What could the others do? Thus there is one thing whose existence is necessary, and if necessary, by the very same argument, eternal. Since it is eternal, it has no cause. Questions about *its* origins are pointless.

What is God if not an infinite and necessarily existing being?

ᘓ

This argument is by no means foolish. It is spacious. It has a certain grandeur. But it remains only as strong as its weakest premise: *If the universe might never have existed, then for sure at some time or other it did not exist.*

When this premise is placed in hot type on cold paper, suspicions arise that it covers an inference that Aquinas cannot support. The steps involved in passing from *I exist* to *I might exist*—they are fine. The additional steps that carry the metaphysician from *I might not exist* to at one time *I did not (or will not) exist*—they are fine too. They are as fine as metaphysical inferences ever get. But to suppose that precisely the same steps carry the *universe* from *it might not exist* to *it did not exist* suggests the fallacy of composition at work, as when the set of turtles is said to be a turtle on the grounds that its members are all turtles. One for all and all for one is *not* a principle of metaphysics. A universe of perishable things is not necessarily perishable. This objection does not by itself close the case. No case in metaphysics or theology is ever closed. But it does indicate that some further argument is needed, and this Aquinas does not provide.

᷒

Let us suppose, then, that the universe passes sedately from the everlasting to the everlasting. It has been there forever and it will be there forever. This is the universe that Einstein championed before he appreciated the explosive nature of Big Bang cosmology, and it is a universe that has always induced a sense of calm in those who contemplate it. If it does not appear to be *the* universe and thus *our* universe, a great many cosmologists in the twentieth century have regarded that as a defect in the plan of creation. A universe of this sort makes a busy, causally imperious God unnecessary; what is worse, it makes him incoherent. A cause must precede its effect, and if the universe is eternal, there was no moment in which God could have brought about the creation of the universe. In a world with so much time, it is odd to think that God—of all people!—would have no time in which to work. The best he could do from the outside would be to barge into the universe occasionally and cause a great deal of commotion.

Nonetheless, an eternal universe leads to a question very similar to the question that Aquinas asked, and it allows us to recapture some of the force of the second cosmological argument without the affliction of a very doubtful premise. The reformation strikes for a deeper level of doubt and perplexity than the original argument and for this reason carries an emotional burden that the original argument lacks.

"If the universe was always there and will always be there, why is it there at all?"

There is no point in answering this question by assuming that our own fond familiar universe *must* exist. With all due respect to the universe, this is an assumption no one wishes to make, because no description that we can offer of the universe suggests that its existence is necessary. But if the universe does not exist necessarily, then plainly it might never have existed at all, *even if* it has existed for all time.

And that is precisely the problem. With the possibility of

inexistence staring it in the face, why *does* the universe exist? To say that universe just *is,* as Stephen Hawking has said, is to reject out of hand any further questions. We *know* that it is. It is right there in plain sight. What philosophers such as ourselves wish to know is *why* it is. It may be that at the end of these inquiries we will answer our own question by saying that the universe exists for no reason whatsoever. *At the end* of these inquiries, and not the beginning.

No matter how cheerfully physicists may endorse this conclusion, it is dreadful.

This is something we know too.

တ

Two arguments are now at work. The first is due to Aquinas.

Its first premise:
> *If the universe is contingent, then at some time it did
> not exist.*

Its second:
> *At that time, it emerged from nothing.*

Its conclusion:
> *This is crazy.*

And the second argument, derived from a mixed salad of philosophical greens of my own devising:

Its first premise:
> *If the universe is contingent, there is no saying
> whether it existed forever. Maybe. Maybe not.*

Its second:

> *If anything might not exist, then it is reasonable to*
> *ask why it does exist.*

Its conclusion:

> *Well, why does it exist? No, I mean really?*

The first argument asks of the universe how it emerged; the second, why it is there.

The first demands a cause; the second, a reason.

Both arguments are inferences to God, but they proceed from different sources in the imagination.

A causally successful God is what He seems: By creating the universe, He has gotten the job done, and if in return He demands a good deal by way of worshipful admiration, who is to gainsay Him?

A God who functions as a reason is occupied with what German metaphysicians might call the foundations of being. He functions as an anchor and so as a refuge.

Both Gods are equally necessary, but the God engaged in anchoring the universe does not necessarily bother himself with its creation. Why should He? The thing has been there forever. His role is otherwise, and it is more fundamental. It is to this lofty and remote deity that the human heart turns when it wishes to assure itself that there is one thing in Being answering to the majesty of I AM THAT I AM.

God is in this sense an answer to the question long posed by metaphysicians: Why is there something rather than nothing?

If anything exists contingently, the second cosmological argument affirms, at least one thing exists necessarily. There *is* something rather than nothing, because at least one part of existence has its origins in what must be. As for the rest of creation, in one way or another it may be allowed to take care of itself. In reaching for a God who exists necessarily, the theologians have covered their most important base.

And the scientists, until now scoffing at the sidelines, what have they to say about all this?

THE HEART OF MATTER

In the early years of the nineteenth century, the English polymath Thomas Young demonstrated that light behaves like a wave. After shining a beam of light through two slits, he observed interference patterns forming on a screen placed behind them. Wave crests met wave crests to form bigger crests; wave troughs met wave troughs to form deeper troughs; and when crests and troughs were not meeting companionably, they interfered with one another in order to extinguish themselves.

What could be simpler? Light is like a wave.

Ah, but on the other hand, Einstein demonstrated in 1905 that in order to explain the photoelectric effect, it was necessary (or at least convenient) to assume that light comprises particles. Send a beam of light toward a metal surface, and electrons pop out. Plainly they pop out because they have been knocked off. To accommodate both popping out and knocking off, Einstein found it necessary to think of light as if it were composed of discrete packets of energy.

What could be simpler? Light is like a particle.

It was not entirely clear how in the matter of *Young v. Einstein,* both men could have been right.

The consortium of physicists who created quantum mechanics in the third decade of the twentieth century—Neils Bohr, Werner Heisenberg, Erwin Schrödinger, Max Born—finessed this problem by declaring *Young v. Einstein* a draw. Light, they argued, is *both* like a wave *and* like a particle, and what is more, it is like a wave and like a particle on the level of individual photons themselves. Photons, physicists came to understand, interfere with *themselves,* and if deep down no one had the slightest idea how to picture autointerference, what physicists were willing to give up was the picture and not the interference.

The finessing required, as one might imagine, a good deal of finesse.

A quantum particle—an electron or a photon, say—is here, and somewhat later, it is there. The old here-and-there, Schrödinger specified in terms of the properties of a wave. It is here where the wave mounts and there where it dips. Passing through two slits, the wave peaks at the left and peaks as well at the right, flowing, as waves tend to do, through both slits at once.

But a wave is intended to track the moving position of a *single* particle, and it is here that the formalism of quantum mechanics commits the physicist to a form of legerdemain that has to this day resisted all attempts at explication. It is one thing to say that a wave may pass through two slits; it is quite another thing to say that a single particle may divide its allegiance in just the same way. Nonetheless, this is just what physicists were forced to say. By now, they say it without a second thought. The particle that could be here or there they represent by a wave that is here and there. If that is where the wave is, the particle enjoys a doubling of its position in space, with each position corresponding to a distinct physical state. Somehow both physical states are real and they are real at the same time. They are, as physicists say, superimposed. They exist together. There is no getting rid of them. Superimposed states are themselves described by the undulation of a wave, which is generally described as a wave packet to signify the extent to which it embodies a variety of different quantum states and so a variety of separate waves. It is Schrödinger's equation that describes the wave packets' undulations.

The formalism of quantum mechanics, physicists at once realized, defeated all efforts to picture the quantum world. If no pictures were available, neither was there a link to common sense. Light is both a wave and a particle, and it is both a wave and a particle at the same time. This conclusion embodies a mystery, one that no subsequent analytical efforts have dissolved. The mystery will not appear entirely unfamiliar to Christians persuaded of the threefold aspect of the deity. If light is a particle and a wave, religious believers might observe, God is a Father, a Son, and a Holy Ghost. This is not an analogy that has captured the allegiance of scientific atheists.

The interpretation of quantum mechanical formalism did little to dispel the mystery it embodied. In 1926, Max Born provided the standard scheme by which the equations of quantum mechanics might be understood. The details are complex, but in a rough-and-ready way, Born suggested that the quantum mechanical waves passing sedately throughout the universe might be understood in terms of the probabilities that they reveal. Thus the amplitude of a wave is a sign that quite likely there is a particle there and so a clue to its position, and the distance between wave peaks is again a sign that quite likely the particle is traveling with a particular momentum. A wave with two peaks rising like the devil's horns might represent a particle dividing its allegiances equally between two slits.

Under Born's interpretation of quantum mechanics, the identity of a particle undergoes further deconstruction. The old here-or-there has long since passed to the new here-*and*-there, but what is here and there is now a matter of chance. Having impossibly divided itself between two slits, a single photon undergoes further demotion to appear in quantum mechanics as the ghost of its position. It *could* be here, it *could* be there, and somehow it *could* be at both places at once.

These divided allegiances come to an end abruptly when an observer, padding in from outside the quantum system, undertakes a measurement. So long as no one is looking, the electron is all things to all men. But let the physicist have a

look, and *boom!* the particle that could be here *and* there becomes here *or* there all over again. The wave packet collapses
into just one of its possibilities. The other quantum states that
it embodies vanish, and they vanish instantaneously.

No one knows why.

Niels Bohr—widely considered to be inscrutable in his
conversation, owing to the particular flavor of his Danish
Grope and Mumble—embraced this interpretation of quantum mechanics, whence its designation as the Copenhagen
interpretation. It has become canonical.

It has not, however, explained the connection between the
quantum realm and the classical realm. "So long as the wave
packet reduction is an essential component [of quantum me-
chanics]," the physicist John Bell observed, "and so long as we
do not know when and how it takes over from the Schrödinger
equation, we do not have an exact and unambiguous formula-
tion of our most fundamental physical theory."

If this is so, why is our most fundamental physical theory
fundamental?

I'm just asking.

SOMETHING FROM NOTHING

Cosmology studies the universe as a whole, and quantum
cosmology brings the apparatus of quantum mechanics
to bear on the whole of the universe. It is the most speculative
of inquiries and it is among the least successful. It seems to
tempt physicists to a certain gracelessness.

Considering the cosmological argument, the physicist Victor Stenger scoffs that it is the "last resort of the theist who seeks to argue for the existence of God from science and finds all his other arguments fail." Sheer chutzpah, if I may use the Greek for cheek. It is *Stenger* who is arguing against the existence of God "from science." The result, as one might expect, is unedifying. "Why," Stenger asks, "is there God rather than nothing?" It is what physicists always ask before they have thought about what they are asking.

If God *must* exist, the question why God *does* exist answers itself. Must is must.

Having rejected Aquinas, Stenger is persuaded that "we can give a plausible *scientific* reason based on our best current knowledge of physics that something is more natural than nothing!" The appeal to what is natural elicits an old urge among physicists to possess the concept of naturalness voluptuously. But it is worth remembering that what is at issue is not whether something is more natural than anything, but *why the universe exists at all*. Naturalness has nothing to do with it.

Oxford's Peters Atkins has attempted to address this issue. "If we are to be honest," he argues, "then we have to accept that science will be able to claim complete success only if it achieves what many might think impossible: accounting for the emergence of everything from absolutely nothing." Atkins does not seem to recognize that when the human mind encounters the thesis that something has emerged from nothing, it is not encountering a question to which any coherent answer exists.

His confidence that a scientific answer must nonetheless be forthcoming needs to be assessed in other terms, possibly those involving clinical self-delusion.

Among physicists, the question of how something emerged from nothing has one decisive effect: It loosens their tongues. "One thing [that] is clear," a physicist writes, "in our framing of questions such as 'How did the Universe get started?' is that the Universe was self-creating. This is not a statement on a 'cause' behind the origin of the Universe, nor is it a statement on a lack of purpose or destiny. It is simply a statement that the Universe was emergent, that the actual Universe probably derived from an indeterminate sea of potentiality that we call the quantum vacuum, whose properties may always remain beyond our current understanding."

It cannot be said that "an indeterminate sea of potentiality" has anything like the *clarifying* effect needed by the discussion, and indeed, except for sheer snobbishness, physicists have offered no reason to prefer this description of the Source of Being to the one offered by Abu al-Hassan al Hashari in ninth-century Baghdad. The various Islamic versions of that indeterminate sea of being he rejected in a spasm of fierce disgust. "We confess," he wrote, "that God is firmly seated on his throne. We confess that God has two hands, *without asking how*. We confess that God has two eyes, *without asking how*. We confess that God has a face."

So long as frank confessions are being undertaken, *I* must confess that a God looking agreeably like me makes precisely as much sense as an "indeterminate sea of potentiality," with the additional advantage that *He* is said to be responsive to prayer.

Having begun with Stenger, I might as well finish him off. Proposing to show how something might emerge from nothing, he introduces "*another universe* [that] existed prior to ours that tunneled through . . . to become our universe. Critics will argue that we have no way of observing such an earlier universe, and so this is not very scientific" (italics added).

This is true. Critics will do just that. Before they do, they will certainly observe that Stenger has completely misunderstood the terms of the problem that he has set himself, and that far from showing how something can arise from nothing, he has shown only that something *might* arise from something else. This is not an observation that has ever evoked a firestorm of controversy.

A man must really know his own limits, as Clint Eastwood observed.

❧

The Sea of Indeterminate Potentiality, and all cognate concepts, belong to a group of physical arguments with two aims. The first is to find a way around the initial singularity of standard Big Bang cosmology. Physicists accept this aim devoutly because the Big Bang singularity strikes an uncomfortably theistic note. Nothing but intellectual mischief can result from leaving that singularity where it is. Who knows what poor ideas religious believers might take from cosmology were they to imagine that in the beginning the universe began?

The second aim is to account for the emergence of the universe in some way that will allow physicists to say with quiet pride that they have gotten the thing to appear from nothing, and especially nothing resembling a deity *or* a singularity.

This is the province of ideas first advanced by Stephen Hawking and James Hartle and later by Hawking, Ian Moss, and Neil Turok. The details may be found in Hawking's best-selling *A Brief History of Time*, a book that was widely considered fascinating by those who did not read it, and incomprehensible by those who did. Their work will seem remarkably familiar to readers who grasp the principle behind pyramid schemes or magical acts in which women disappear into a box only to emerge as tigers shortly thereafter.

Quantum mechanics of the old-fashioned kind assesses the behavior of particles, chiefly by showing that particles are not particles at all but a kind of probabilistic smear. In quantum cosmology, the particles are gone. Gone as well is the classical form of Schrödinger's equation, though its domestic companion, a wave function taking *universes* as its objects (more or less), also operates in terms of probabilities.

Quantum cosmology dispenses with the Copenhagen interpretation's queer distinction between the quantum world and the classical world, wherein the electron belongs to the quantum world, the physicist to the classical world. There are no classical physicists loitering about quantum cosmology, and no classical world either. It is quantum mechanics all the way down, and, of course, all the way up as well.

Now, when Schrödinger first came to appreciate the mysteries of quantum theory, he devised a thought experiment to explain his own perplexity. Imagine that a cat has been placed in a sealed container, together with a device that *if* it goes off will kill it—a revolver, say, or some sort of radioactive pellet. *Whether* the device goes off is a matter of chance. So long as no one is looking, the cat exists in a superposition of quantum states, at once half dead (the gun might fire) and half alive (it might not). As soon as an observer peeks into the box, that superposition gives way. That cat is either dead or alive and there are no two ways about it. Schrödinger thought the idea of a cat both alive and dead intellectually discouraging.

Schrödinger's cat is a part of the mythology of quantum theory, and according to the Copenhagen interpretation, it is there for the count, because no one can imagine how to get rid of the poor creature.

For this reason any number of physicists have endeavored to get rid of the Copenhagen interpretation instead. In 1957, Hugh Everett III, a young physicist at Princeton, argued in his Ph.D. dissertation that the collapse of the wave function could be explained on the assumption that reality somehow contains far more worlds than previously imagined. Where an observer in classical quantum theory would occupy himself in collapsing what we may fondly recall as the good, old-fashioned wave function, according to the many-worlds interpretation, at precisely the moment a measurement is made, the universe branches into two or more universes. The cat who was half dead and half alive gives rise to two separate universes, one containing a cat who is dead, the other containing a cat who is alive. The new universes cluttering up creation embody the quantum states that were previously in a state of quantum superposition.

The many-worlds interpretation of quantum mechanics is rather like the incarnation. It appeals to those who believe in it, and it rewards belief in proportion to which belief is sincere.

෧

The wave function of the universe is designed to represent the behavior of the universe—*all of it*. It floats in the void—these metaphors are inescapable—and passes judgment on universes. Some are probable, others likely, and still others a very bad bet. Nonetheless, the wave function of the universe cannot be seen, measured, assessed, or tested. It is purely a theoretical artifact. Physicists have found it remarkably easy to pass from speculation *about* the wave function of the universe to the conviction that there *is* a wave function of the universe. This is nothing more than an endearing human weakness. Less endearing by far is their sullen contempt toward religious argument when it is engaged in precisely the same attempt to reach by speculation what cannot be grasped in any other way.

By itself, the wave function of the universe can do little to advance the double agenda of quantum cosmology: to get rid of the initial singularity of Big Bang cosmology, and to show how the universe emerged from nothing much or nothing at all. It is a necessary piece of equipment, like the mountain climber's rope.

What the physicist requires to get climbing is a readjustment of our traditional physical notions of time, a way of giving it a new look. The new look is necessary because, as Stephen Hawking and Roger Penrose demonstrated in the mid-1960s, the Big Bang singularity is simply unavoidable. Within general relativity, time has an unvarying direction. If a man is going down toward the Big Bang, it is one thing *before* another, and if he is coming up from the Big Bang, one thing *after* another. This is a feature of the real number system itself. It cannot be changed. Within quantum cosmology, however, time has been altered. Very much like a physician who proposes to cure his patient's infection by infecting him with another affliction, Hawking suggested that in going down toward the Big Bang, one mathematical regime (that of the real numbers) would somehow give way to another (that of the imaginary numbers).

It was the use of the word *imaginary* in this context that gave his ideas their air of pontifical mystification. How can numbers be imaginary? They cannot be. Hawking was simply appealing to the complex numbers, and these are perfectly well-defined mathematical objects. They correspond more or less to pairs of points in the plane.

The complex numbers have one outstanding advantage:

They are not ordered. They do not *go* anywhere. If time is measured by the complex numbers, there is no *before* at work and no worries at all about winding up at the Big Bang singularity. Thus in Hawking's scheme, at the point in which the regime of the real numbers gives way, the complex regime takes over. As the physicist descends toward the place formerly known as the Big Bang singularity, time smoothly executes a transformation all its own, the region around the tip becoming gently curved, so that the cone ends in a pendulous sac. There is now a moment corresponding to the magician's withdrawal of a handkerchief from his sleeve: *The Big Bang singularity has disappeared!*

It is just gone.

Within the sac, the physicist cannot see or otherwise determine a before *before* his last before. He is adrift in a directionless borough of space and time.

It is very much like Brooklyn, one reason that the early universe (and everyone else) was so eager to get out of there.

CAN THEY GET AWAY WITH IT?

In commenting on the scenario described by Hawking and his colleagues, Roger Penrose, writing in *The Road to Reality*, offered his opinion that their theories were remarkably elegant. It was a gracious remark. A far more natural reaction would be to ask, "Can they really get away with that?" From a technical point of view the answer is yes. They have the mathematical means. In going down, one version of space and time gives way. Another becomes ascendant. A fog of sorts begins to cover everything. It disappears on coming up. In between the going down and the coming up, the original Big Bang singularity has vanished.

When scholars persuaded of the essential inerrancy of the Bible attempt to reconcile the Book of Genesis with contemporary estimates of the age of the cosmos, they do so by changing the time mentioned in the Bible and so altering its nature. These efforts are not necessarily foolish. Often there is real ingenuity required, and no little physical competence. The physicist Gerard Schroeder is convinced that the Hebrew Bible provides a stunning insight into the cosmos of creation, and he has traveled the world in an effort to present his views. They have not been well received by physicists, who in their retirement often enjoy writing critical assessments of biblical scholarship, a vocation that allows them to demonstrate their knowledge without ever defending it. The gravamen of their concerns lies less with the plausibility of various schemes than with their motivation. And that is frankly and honestly in the service of a religious agenda.

And Hawking?

The question is leading, Your Honor, I know that, but look *where* it is leading.

Never mind looking, if you are otherwise occupied. I'll point out the place myself. It is leading to a place that anyone who follows human thought should find familiar. Arguments follow from assumptions, and assumptions follow from beliefs, and very rarely—perhaps never—do beliefs reflect an agenda determined entirely by the facts. No less than the doctrines of religious belief, the doctrines of quantum cosmology are what they seem: biased, partial, inconclusive, and largely in the service of passionate but unexamined conviction.

There is no surprise in any of this, and if there is, there should not be.

With the Big Bang singularity removed from sight, there remains the second part of quantum cosmology's two-part agenda, and that is to provide a scenario for the emergence of the universe—our own universe, that is, now demoted in grandeur from *the* universe to one among many.

The argument that Hawking has offered may be conveyed by question-and-answer, as in the Catholic catechism.

A Catechism of Quantum Cosmology

Q: From what did our universe evolve?

A: *Our universe evolved from a much smaller, much emptier mini-universe. You may think of it as an egg.*

Q: What was the smaller, emptier universe like?

A: *It was a four-dimensional sphere with nothing much inside it. You may think of that as weird.*

Q: How can a sphere have four dimensions?

A: *A sphere may have four dimensions if it has one more*

dimension than a three-dimensional sphere. You may think of that as obvious.

Q: Does the smaller, emptier universe have a name?

A: *The smaller, emptier universe is called a de Sitter universe. You may think of that as about time someone paid attention to de Sitter.*

Q: Is there anything else I should know about the smaller, emptier universe?

A: *Yes. It represents a solution to Einstein's field equations. You may think of that as a good thing.*

Q: Where was that smaller, emptier universe or egg?

A: *It was in the place where space as we know it did not exist. You may think of it as a sac.*

Q: When was it there?

A: *It was there at the time when time as we know it did not exist. You may think of it as a mystery.*

Q: Where did the egg come from?

A: *The egg did not actually come from anywhere. You may think of this as astonishing.*

Q: If the egg did not come from anywhere, how did it get there?

A: *The egg got there because the wave function of the universe said it was probable. You may think of this as a done deal.*

Q: How did our universe evolve from the egg?

A: *It evolved by inflating itself up from its sac to become the universe in which we now find ourselves. You may think of that as just one of those things.*

This catechism, I should add, is not a parody of quantum cosmology. It *is* quantum cosmology.

Readers lacking faith will, I imagine, wish to know something more about its crucial step, and that is the emergence of a mini-universe from nothing at all. They will be disappointed to learn that insofar as the mini-universe is actual, it did not emerge from nothing, and insofar as it is possible, it did not emerge at all. What can be said about the mini-universe according to either interpretation is that Hawking has designated it as probable because he has assumed that it *is* probable. He has done this by restricting the wave function of the universe to just those universes that coincide with the de Sitter universe at their boundaries. This coincidence is all that is needed to produce the desired results. The wave function of the universe and the de Sitter mini-universe are made for each other. The subsequent computations indicate the obvious: The universe most likely to be found down there in the sac of time is just the universe Hawking assumed would be found down there. If what Hawking has described is not quite a circle in thought, it does appear to suggest an oblate spheroid.

The result is guaranteed—*one hunnerd* percent, as used-car salesmen say.

⟿

Among philosophers concerned to promote atheism, satisfaction in Hawking's conclusion has been considerable. Witness Quentin Smith: "Now Stephen Hawking's theory dissolves any worries about how the universe could begin to exist uncaused." Smith is so pleased by the conclusion of Hawking's argument that he has not concerned himself overmuch with its premises. Or with its reasoning.

While Hawking's scheme has since its inception been the subject of many technical and philosophical criticisms, disputes have been, I must say, disappointingly courteous. Unlike particle physicists, whose natural level of aggression compares favorably with that of the timber wolf, cosmologists are often languid in argument, and they attend to the deficiencies of one another's work with the studied elegance of men who keep silk handkerchiefs in their sleeve.

In 1984, Alexander Vilenkin published a paper adverting to the creation of the universe out of nothing. According to his view, the universe *tunneled* its way into becoming a de Sitter universe. Twenty years later, he was moved in a paper entitled "Quantum Cosmology and Eternal Inflation" to ask whether his original paper might not have been his "greatest mistake." Clearly he was not in this regard worried about an embarrassment of riches. On more sober reflection, he decided the point in his favor. At the conclusion of his paper, he observed that "sadly, quantum cosmology is not likely to become an observational science."

Correct. Quantum cosmology is a branch of mathematical metaphysics. It provides no cause for the emergence of the universe, and so does not answer the first cosmological question, and it offers no reason for the existence of the universe, and so does not address the second. If the mystification induced by its modest mathematics were removed from the subject, what remains would not appear appreciably different in kind from various creation myths in which the origin of the universe is attributed to sexual congress between primordial deities.

6

A Put-up Job

"THOUSANDS HAVE lived without love," W. H. Auden observed, "not one without water." Love is important; water is *necessary*. If water is necessary, so, too, a great many other things. In a paper entitled "Large Number Coincidences and the Anthropic Principle in Cosmology," published in 1974, the physicist Brandon Carter observed that many physical properties of the universe appeared fine-tuned to permit the appearance of living systems.

What a lucky break—*things have just worked out.*

What an odd turn of phrase—*fine-tuned.*

What an unexpected word—*permit.*

Whether lucky, odd, or unexpected, the facts are clear. The

cosmological constant is a number controlling the expansion of the universe. If it were negative, the universe would appear doomed to contract in upon itself, and if positive, equally doomed to expand out from itself. Like the rest of us, the universe is apparently doomed no matter what it does. And here is the odd point: If the cosmological constant were larger than it is, the universe would have expanded too quickly, and if smaller, it would have collapsed too early, to permit the appearance of living systems. Very similar observations have been made with respect to the fine-structure constant, the ratio of neutrons to protons, the ratio of the electromagnetic force to the gravitational force, even the speed of light.

Why stop? The second law of thermodynamics affirms that, in a general way, things are running down. The entropy of the universe is everywhere increasing. But if things are running down, what are they running down *from*? This is the question that physicist and mathematician Roger Penrose asked. And considering the rundown, he could only conclude that the runup was an initial state of the universe whose entropy was very, very low and so very finely tuned.

Who ordered *that*?

"Scientists," the physicist Paul Davies has observed, "are slowly waking up to an inconvenient truth—the universe looks suspiciously like a fix. The issue concerns the very laws of nature themselves. For 40 years, physicists and cosmologists have been quietly collecting examples of all too convenient 'coincidences' and special features in the underlying laws of the universe that seem to be necessary in order for life, and hence conscious beings, to exist. Change any one of them and the consequences would be lethal."

These arguments are very much of a piece with those that Fred Hoyle advanced after studying the resonances of carbon during nucleosynthesis. "The universe," he grumbled afterward, "looks like a put-up job." An atheist, Hoyle did not care to consider who might have put the job up, and when pressed, he took refuge in the hypothesis that aliens were at fault. In this master stroke he was joined later by Francis Crick. When aliens are dropped from the argument, there remains a very intriguing question: *Why* do the constants and parameters of theoretical physics obey such tight constraints?

If this is one question, it leads at once to another. The laws of nature are what they are. They are fundamental. But why are they true? Why do material objects attract one another throughout the universe with a kind of brute and aching inevitability? Why is space-and-time curved by the presence of matter? Why is the electron charged?

Why? Yes, *why*?

An appeal to still further physical laws is, of course, ruled out on the grounds that the fundamental laws of nature are *fundamental*. An appeal to logic is unavailing. The laws of nature do not seem to be logical truths. The laws of nature must be intrinsically rich enough to specify the panorama of the universe, and the universe is anything but simple. As Newton remarks, "Blind metaphysical necessity, which is certainly the same always and everywhere, could produce no variety of things."

If the laws of nature are neither necessary nor simple, why, then, *are* they true?

Questions about the parameters and laws of physics form a single insistent question in thought: *Why* are things as they are when *what* they are seems anything but arbitrary?

One answer is obvious. It is the one that theologians have always offered: *The universe looks like a put-up job because it is a put-up job.* That this answer is obvious is no reason to think it false. Nonetheless, the answer that common sense might suggest is deficient in one respect: It is emotionally unacceptable because a universe that looks like a put-up job puts off a great many physicists.

They have thus made every effort to find an alternative. Did you imagine that science was a *disinterested* pursuit of the truth?

Well, you were wrong.

APOTHEOSIS IN THE STANDARD MODEL

At the beginning of the 1960s, physicists understood that there were four forces in play in the material world: the force of gravitation, the electromagnetic force, and the weak and strong nuclear forces. They had in addition come into possession of a remarkably large number of elementary particles, so many that Enrico Fermi complained that had he wished to memorize their names, he would have become a botanist.

Thirteen years later, three of the four forces and virtually all of the elementary particles had been successfully classified, and the forces partially explained because partially unified. This is the triumph of the Standard Model.

It is a model comprising three parts. The first is quantum electrodynamics, which offers a successful quantum theory of the electromagnetic field, one satisfying principles of both quantum mechanics and special relativity. Quantum electrodynamics was completed in the late 1940s by Richard Feynman, Julian Schwinger, and Sin-Itiro Tomonaga; and because it describes electromagnetic phenomena—light, electricity, magnetism—it retains a vivid connection with the world of daily life in which computer chips and electric toasters hum in accordance with its laws. Without it, we would all be lost, or at best, inconvenienced.

The second part of the Standard Model, Steven Weinberg, Sheldon Glashow, and Abdus Salaam created in their electroweak theory. As the name might indicate, their theory unified the weak nuclear force and the electromagnetic force. By showing that, deep down, two forces were really one, Weinberg, Glashow, and Salaam demonstrated that when properly seen, the weak nuclear force and the electromagnetic force were manifestations of some ancient primordial form of unity. In the world as it is, of course, very little of this unity is left. The weak nuclear force and the electromagnetic force are today distinct. To see things as they really are, it is necessary to see things as they really were. The time when things really were unified occurred shortly after the Big Bang. To account for the fact that in the world as it *is* observed, the weak force and the electromagnetic force are distinct, Weinberg, Glashow, and Salaam appealed to the audacious idea that what physicists could today see of the weak and electromagnetic forces represented nothing more than a form of broken symmetry, as when couples remember how happy they once were amid the shambles of their discontent.

There is finally quantum chromodynamics, which provides a theory of the strong nuclear force. In 1954, C. N. Yang and Robert Mills outlined a daring generalization of quantum electrodynamics. Their paper described a new physical theory. It also predicted the existence of particles that no experiment had revealed and strange new symmetries.

With the proliferation of quarks and their varieties in the 1960s, new particles and symmetries did emerge, and they proved to be precisely those that would allow a Yang-Mills theory to take charge of the strong nuclear force and give it direction and a general shaping-up.

There followed a decisive step, the last. Experiments had indicated that in some bizarre fashion, particles bound by the strong nuclear force behaved in ways quite unlike particles governed by the weak nuclear force—or any other force, for that matter. Their interactions seemed to grow stronger as the distance between them increased, almost as if they were being held together by a rubber band that remained flaccid at short distances and tense at longer distances. Many marriages are like this. In the early 1970s, David Gross, H. David Politzer, and Frank Wilczek discovered in their theory of asymptotic freedom that this was an expected consequence of a Yang-Mills theory of the strong nuclear force.

The Standard Model was complete.

⌖

If the Standard Model is a triumph, is not one that is unalloyed. The Standard Model cannot explain the transition from the elementary particles to states of matter in which the elementary particles are bound to one another and so form complex structures. It is in this sense incomplete.

The Standard Model is not only incomplete but arbitrary. Like any physical theory, it contains a good many numerical parameters—at least twenty-one. These designate specific numerical properties of the model. These cannot be derived from the theory. Physicists thus find themselves very much in the position of a master couturier obliged to allow one of his finest creations to appear on the runway with its basting lines and tacking pins still affixed.

Above all, the Standard Model is inadequate because it does not incorporate the force of gravity. General relativity stands apart. The two great theories of the twentieth century have not been reconciled. They invoke different languages, different ideas, and different techniques of calculations. The great technical triumphs that made the Standard Model a success are with respect to general relativity unavailing because ineffective. General relativity and quantum mechanics resemble two aging matadors facing the bull of nature, the both of them retiring flustered after a number of halfhearted veronicas and ineffective passes.

The bull is still there, snorting through velvet nostrils. He does not seem the least bit fatigued.

OVERFLOW INTO STRINGS

For the past quarter-century, a very substantial portion of the community of mathematical physicists has been engaged in work on a subject known as string theory. The effort has consumed the best minds of a generation.

Whereupon the inevitable, *Wait a minute, strings?*

Yes, strings. A string is just what its name suggests. It is a wiggling one-dimensional object, something like a garden hose although somewhat smaller, and extended in length but not width. Strings can be straight, they can be curved, they can join with themselves to form loops, and what is more, since they are strings, they can vibrate under tension.

The idea has had a tremendous unifying power, suggesting that nature's elementary particles could be recovered from *one* fundamental object vibrating in various ways. In place of the very complicated system of precisely adjusted forces and parameters characteristic of the Standard Model, string theory pointed to two, and only two, fundamental constraints: The first reflected the string's tension, and so served as the key to its powers of creation; and the second, its coupling constant, the measure of how likely it was to break into two.

Nothing more was needed. This was widely considered a very fine thing.

There followed an illumination, one that lit up all of particle physics. Working very much in isolation, the physicists Joël Scherk and John Schwarz observed that string theory, no matter how manipulated, seemed to predict the existence of a new particle, something like the photon. This seemed uncalled for and therefore unwanted, until physicists realized that amid all those twitching strings, a particle had appeared conveying the force of gravity. For the first time, a fundamental theory in particle physics incorporated a long-missing force. A grand unification seemed to be at hand, one involving *all* nature's forces. No theory could be more final—or more desired—than this.

From that moment on, a number of physicists had the rarest of all experiences: They came to believe that they could hear Nature herself knocking at their door.

୶

In the years that followed—roughly from the late 1970s until the present—string theory expanded and grew great. Difficulties appeared and were surmounted, whereupon new difficulties appeared. Physicists were obligated to undertake very difficult calculations with respect to a theory that they did not completely understand. Their work revealed strange coincidences and tantalizing suggestions of a deeper form of unity. By the early part of the twenty-first century, they could look back on two string theoretic revolutions, and while both advanced the cause, neither brought the goal of a single, clearly stated final theory within reach.

The reaction, although slow in coming, was also inevitable. String theory was criticized in the popular press by a distinguished theoretical physicist and a mathematician. In *The Trouble with Physics*, written by Lee Smolin, and *Not Even Wrong*, by Peter Woit, string theory was examined with some sympathy and found wanting. Neither author could find a theory in the place where theoreticians said a theory should be, and both authors noted with some asperity that string theory had no apparent connections to experiment and that none were in prospect. Woit went so far as to observe that the mathematical structure on which the theory rested, far from being a thing of great elegance, was the most horrible thing he had ever seen.

꿈

Whatever their other merits, all string theories are characterized by an embarrassing dimensional overflow. Some versions of string theory require twenty-six dimensions; others, ten; and still others, eleven. Our own universe contains only three or four, but in any case, no more than a handful. It is one thing to consider higher dimensions as mathematical artifacts. Mathematicians have no difficulty in dealing with an *infinite* dimensional space. They do it all the time. But the extra dimensions of string theory are not purely mathematical. They are within string theory quite real, if only because they have useful work they must do. If real, those extra dimensions are nonetheless invisible. As one might easily imagine, the conflict between the demands of theory—*Get me those extra dimensions*—and the constraints of common sense—*No extra dimensions here, Boss, and we looked*—was not easily resolved.

In the end, string theorists argued that the extra dimensions of their theory were buried somewhere. At each point in space and time, they conjectured, *there* one would find a tiny geometrical object known as a Calabi-Yau manifold, and curled up within, *there* one would find the extra dimensions of string theory itself.

It was an idea that possessed every advantage except clarity, elegance, and a demonstrated connection to reality.

With extra dimensions buried, stable solutions emerged from the equations of string theory, just as the physicists had hoped. They were not, unfortunately, unique. There were thousands of them, and each led to a different version of the theory, a point in an enormous space of possibilities, a landscape of a sort never seen before, a place where each point seemed to embody a different scheme of physical thought, and so a different universe governed by the scheme. In its appearance in various popular journals, the mutant thing was depicted as a gigantic set of bubbles floating in space, our own universe a dimpled dot lost somewhere amid that infernally expanding froth.

FLIGHT INTO THE FANTASTIC

String theory confronted the community of particle physicists with an exquisite dilemma. A theory that initially seemed too *good* to be true had by the late 1990s seemed too good to be *true*. This was widely considered monstrously unjust.

If string theory did not uniquely describe one universe, physicists reasoned, the fault lay with our universe: It was not man enough to handle so promiscuous a theory. One universe having proved inadequate, more would be required. Endeavoring to unify the forces of nature, physicists determined to multiply the universes in which they were satisfied. Very few physicists appreciated the irony involved in pursuing the first ambition by embracing the second. The physicist Leonard Susskind thus claimed that "the narrow 20th-century view of a unique universe, about ten billion years old and ten billion light years across with a unique set of physical laws, is giving way to something far bigger and pregnant with new possibilities."

Far bigger? And pregnant too? In service to this idea, Susskind wrote that "physicists and cosmologists are coming to see our ten billion light years as an infinitesimal pocket of a stupendous megaverse." On reflection, Susskind came to understand that the word *megaverse* carried negative class associations, as in mega-blockbuster (a movie no one wishes to see) or mega-mall (a place no one wishes to go), whereupon he renamed the megaverse "the Landscape."

The Landscape at once suggested the radical changes to come. "Theoretical physicists," Susskind wrote, "are proposing theories which demote our ordinary laws of nature to a tiny corner of a gigantic landscape of mathematical possibilities."

Each of the versions of string theory is thus free to find its home in some particular universe. Like Odysseus worshipping in foreign temples, there is a universe in which a very large cosmological constant is made to feel welcome. The MIT physicist Max Tegmark is persuaded that this is so, and if in some universe he is persuaded that it is not so, he has learned to accept the emotional incoherence that would trouble others with equanimity.

However named, the Landscape was a provocative, and even a revolutionary idea. Physicists appreciate revolutions for obvious reasons: They stir the blood. "We may be at a new turning point, a radical change in what we accept as a legitimate foundation for a physical theory," Steven Weinberg wrote. It would be hard to imagine a doctrine more radical than the thesis that when it comes to universes, there are a great many of them. At a conference on string theory held in 2005, Weinberg buoyantly indicated that he was prepared to welcome his new insect overlords.

An informal poll indicated that the audience of physicists rejected his views by a margin of four to one.

"We win some and we lose some," Weinberg remarked equably.

֍

The Landscape is a new idea in physical thought, but it is not a new idea. Philosophers have long found the restriction of their thoughts to just one universe burdensome. In the late 1960s, David Lewis assigned possible worlds ontological benefits previously reserved to worlds that are real. In some possible world, Lewis argued, Julius Caesar is very much alive. He is endeavoring to cross the Hudson instead of the Rubicon, and fuming, no doubt, at the delays before the toll booth on the George Washington Bridge. It is just as parochial to reject this world as unreal, Lewis argued, as it would be to reject Chicago because it cannot be seen from New York. Lewis argued brilliantly for this idea, known as modal realism. The absurdity of the resulting view was not an impediment to his satisfaction. Or to mine, needless to say.

Quantum mechanics has also invited the promotion of possible worlds to the ontological Big Time, as readers may remember from chapter 5, where dead-cat universes proliferated alongside universes containing live cats.

During the 1980s, the physicist Alan Guth argued that the early universe was characterized by a period of exponential inflation. Very soon after it blew up in the first place, it blew up again. When suitably blown up, it stopped blowing up. The Stanford physicist André Linde carried this idea a step further in his theory of eternal chaotic inflation. Universes are blowing up all over the place. They cannot stop themselves.

When string theorists talk about the Landscape, they are among friends. If their friends were willing to believe in anything, string theorists, having so lately consorted with twenty-six dimensions, are hardly in a position to complain.

There is no need to turn to such esoteric doctrines to capture the underlying current of thought that animates the Landscape. It is simply the claim that given sufficiently many universes, what is true here need not be true there, and vice versa. This thesis has been current in every college classroom for at least fifty years. It arises spontaneously in discussion, like soap bubbles in water. It is expressed in the same way and often by the same stolid, heavy-thighed undergraduate—a Mr. Waldburg, in my case and class.

After raising his hand with the air of a man compelled to observe the obvious, he has this to say: *There are no absolute truths.*

Waldburg, meet Weinberg.

THE SURE THING

Although initiated as a whim, the Landscape has been welcomed by string theorists as a deliverance. Whether string theory is rescued by the Landscape is relatively a trivial matter. Theories come and go, and if this one goes, another is sure to come. The Landscape has acquired a life of its own because it is addressed to issues that arise *whatever* the theory *whenever* it arrives. If science, as the French mathematician René Thom once remarked, is an attempt to reduce the arbitrariness

of our descriptions, then every theory short of one that is logically necessary must in the end provoke the same two questions: Why are its numerical parameters *as* they are? And why
are its assumptions *what* they are?

The Landscape provides a generic answer. It is all-purpose
in its intent. It works no matter the theory. And it works by
means of the simple principle that by multiplying universes,
the Landscape dissolves improbabilities. To the question *What
are the odds?* the Landscape provides the invigorating answer
that it hardly matters. If the fine-structure constant has in our
universe one value, in some other universe it has another
value. Given sufficiently many universes, things improbable in
one must from the perspective of them all appear certain.

The same reasoning applies to questions about the laws
of nature. Why is Newton's universal law of gravitation true?
No need to ask. In another universe, it is not.

The Big Fix has by this maneuver been supplanted by the
Sure Thing.

৵

As one half of the flight into the fantastic, the Landscape does
what it can, and what it does, it does very well. It dilutes the
acrid acid of improbability. But as philosophers and physicists
at once observed, the Landscape offers a general solution to
what is, in fact, a particular problem. The multiplication of
universes establishes that in some universe, the fine-structure

constant will take *any* designated value. It is a Sure Thing. Nonetheless, the Sure Thing establishes only that life's lucky numbers will sooner or later turn up somewhere or other.

And yet they have turned up *here*, just where we need them the most. Requiring certain amenities, we find ourselves in a universe in which they have been liberally supplied. This may not be a paradox in thought, but surely it seems a suspiciously good deal. We might well have found ourselves in a far less agreeable universe, one in which none of life's lucky numbers were tuned to their sweet spot.

And where would we have been then?

The Landscape now works hand in glove with a second radical idea in physical thought. In the same paper in which he drew attention to the question of fine tuning, Brandon Carter observed that "the universe must be such as to admit the creation of observers within it at some stage." Such is the Anthropic Principle, or, at least, one of them, since the principle now comes in a variety of forms and flavors. It consists, when analyzed, of two quite separate claims.

The first is a matter of common sense. If the universe had not admitted the creation of observers at some stage, why, then, we would not be here.

The second is a claim about the facts of life. If we are surprised by a universe in which we have been given what we need, some part of that surprise, Carter argued, represents a form of bad faith. If the necessities of life are necessary, they must be inevitable. And if inevitable, whence the surprise?

The simple fact that we are *where* we are is sufficient to explain *why* we have *what* we have.

What more could anyone ask?

The question why the ultimate laws of nature are true, and why its numerical parameters have the value that they do, now admits of a two-part response. The first is provided by the Landscape. Neither the numbers nor the laws represent anything improbable. And the second by the Anthropic Principle: If they were false, or if they had different values, where would you be?

Nowhere, right?

And yet here you are.

What *did* you expect?

IF EVERYTHING GOES

The great difficulty with the Landscape and the Anthropic Principle is that physicists prepared to welcome these ideas had no way in which to control them, while physicists prepared to reject them had no way in which to avoid them. In a stimulating paper entitled "Multiverses and Physical Cosmology," the distinguished cosmologists G.F.R. Ellis, U. Kirchner, and W. R. Stoeger considered the idea that in the Landscape anything goes because everything is possible. "In some universes," they write, "there will be a fundamental unification of physics expressible in a basic 'theory of everything,' in others this will not be so."

But having advanced this conjecture, Ellis, Kirchner, and

Stoeger have neglected to tell us whether *it* is true across the Landscape. If so, then not everything goes; and if not, how could it be of interest?

This is, to be sure, something that Ellis, Kirchner, and Stoeger recognize. At the beginning of their essay, they observe that "the very existence of [the Landscape] is based on an assumed set of laws . . . which all universes . . . have in common." It is only later in their essay that they forget what they have written.

I know just how it is, fellas. I can never remember where I left my keys.

The speed with which a commitment to the Landscape ends in incoherence, while it is alarming, is not unexpected. "Any scientist," Steven Weinberg writes in defending his endorsement of anthropic reasoning, "must live in a part of the landscape where physical parameters take values suitable for the appearance of life and its evolution into scientists." To say that portions of the Landscape are "suitable" for the appearance of life is to say that it is *there* that life is *possible*. But if life is possible there, it is not possible elsewhere. Human beings could not, presumably, investigate the universe from the interior of the sun. It is too warm and entirely too gassy. If life is not possible elsewhere, then it is *necessarily* impossible elsewhere. But what might justify this powerful claim if not some physical principle true *everywhere*? If a principle about life is general throughout the Landscape, this would seem to make purely local matters of biology supreme matters of physical thought. This assigns to living systems a degree of cosmic importance that only theologians suspected they possessed.

Given issues such as these, it is at least possible to wonder whether the Landscape and the Anthropic Principle are contrivances in just the sense that Ptolemaic epicycles were contrivances. The Landscape has, after all, been brought into existence by assumption. It cannot be observed. It embodies an article of faith, and like so much that is a matter of faith, the Landscape is vulnerable to the sadness of doubt. There are by now thousands of professional papers about the Landscape, and reading even a handful makes for the uneasy conviction that were physicists to stop writing about the place, the Landscape, like Atlantis, would stop existing—just like that.

This cannot be said of the sun.

When physicists come to defend the Landscape, they use language more commonly heard among biologists. Lee Smolin has argued that deep down there is little evidence in favor of string theory, and even less in favor of the Landscape. *So, what of it?* Leonard Susskind responded: "The level of confidence that string theorists have for their theory is based on a web of interconnected pieces of evidence that is so compelling that genuine mathematicians have no doubt about its validity."

Sentiments of this sort must be appreciated for their speculative inventiveness, if nothing else. Evidence so compelling that no part of it need be produced is not evidence at all. The thesis that a scientific theory represents a "web of interconnected pieces" describes with some economy of effect the *Summa Theologica* of Thomas Aquinas. Or a house of cards.

Very basic physical questions about the Landscape have yet to be answered. On the one hand, there are a very large number of physical theories. They represent a spectrum of possibilities, an immersion into what laws might be true, and what numerical parameters might be in control of things. On the other hand, there are the universes in which they are satisfied, strange, remote, distant, unrecoverable. Physicists very often write as if in the crucible of creation, universes were forever pullulating, red-eyed and throbbing with energy. Perhaps this is so. Who am I to say? But what is left unexplained on these stirring metaphysical accounts is the relationship between those numberless theories and those numberless universes. Just how does a theory get hold of a universe in order to control its birth, formation, and development?

It must do that, because in the end this is just what a theory does, and if it does not do it, then nothing in the Landscape is explained by anything.

But this once again returns the discussion to the point at which it began. If there are such overall principles in charge of the Landscape, why are they true?

Questions such as this reflect in the end a single point of intellectual incoherence. The thesis that there are no absolute truths—is it an absolute truth? If it is, then some truths are absolute after all, and if some are, why not others? If it is not, just why should we pay it any mind, since its claims on our attention will vary according to circumstance?

৻ঔ৾

As a physical claim, the Anthropic Principle hardly seems to enjoy the same authority as the conservation of energy. It is in one sense trivial. We see what we can. But efforts to move the principle from the place in which platitudes congregate have not been entirely successful. Can we really explain the necessities of life by the fact that we are enjoying them? In I Kings of the Hebrew Bible, the prophet Elijah, lost in the desert and without food or water, sat underneath a juniper tree and waited for death. An angel appeared, offering him refreshment. What Elijah took, he of course needed, and since he needed what he took, what he took was sufficient to account for his survival. Biblical commentators have wisely refrained from explaining the angel's appearance on these grounds. The angel, they observed, was sent to Elijah by God. That is the proper explanation for its appearance. No matter the extent to which we need the laws and parameters of the physical world to be as they are, that by itself cannot explain the fact that they are as they are.

৻ঔ৾

It is odd that men who as a group are united by their conviction that religious beliefs are very primitive should find themselves disputing matters more commonly discussed in the Alpha Phi Alpha keg room. It is so nonetheless, a point that the brothers find hardly surprising. Discussions on various Internet postings are endless. Often they contain an eerie mixture of technical sophistication and philosophical incompetence. Or the other way around. The willingness of physical scientists to explore such strategies in thought might suggest to a perceptive psychoanalyst a desire not so much to discover a new idea as to avoid an old one.

Such things happen. And they happen even in mathematical physics.

Received wisdom has it that lacking access to the mysteries of science, men and women accept instead the mysteries of faith. This diagnosis is very often expressed in terms of evolutionary theory. The human brain is an instrument shaped by selection for survival, and it is only natural, considering the problems they faced many years ago, that anxious men and women should have turned toward elaborate theological speculation. What better hedge against fearsome predators or an uncertain food supply than the Immaculate Conception or the revelations of the *Gematria*? As general relativity or quantum field theory become more widely known, human gullibility will decline.

This is not a view of things that a close study of string theory, the Landscape, or the Anthropic Principle tends to support.

GOD, LOGIC, NOTHING

Joel Primack, a cosmologist at the University of California, Santa Cruz, once posed an interesting question to the physicist Neil Turok: "What is it that makes the electrons continue to follow the laws."

Turok was surprised by the question; he recognized its force. Something seems to compel physical objects to obey the laws of nature, and what makes this observation odd is just that neither compulsion nor obedience are physical ideas.

Medieval theologians understood the question, and they appreciated its power. They offered in response the answer that to their way of thinking made intuitive sense: *Deus est ubique conservans mumdum.* God is everywhere conserving the world.

It is *God* that makes the electron follow *His* laws.

Albert Einstein understood the question as well. His deepest intellectual urge, he remarked, was to know whether God had any *choice* in the creation of the universe. If He did, then the laws of nature are as they are in virtue of His choice. If He did not, then the laws of nature must be necessary, their binding sense of obligation imposed on the cosmos in virtue of their form. The electron thus follows the laws of nature because it cannot do anything else.

It is *logic* that makes the electron follow *its* laws.

And Brandon Carter, Leonard Susskind, and Steven Weinberg understand the question as well. Their answer is the

Landscape and the Anthropic Principle. There are universes in which the electron continues to follow some law, and those in which it does not. In a Landscape in which anything is possible, nothing is necessary. In a universe in which nothing is necessary, anything is possible.

It is *nothing* that makes the electron follow *any* laws.

Which, then, is it to be: God, logic, or nothing?

This is the question to which all discussions of the Landscape and the Anthropic Principle are tending, and because the same question can be raised with respect to moral thought, it is a question with an immense and disturbing intellectual power.

For scientific atheists, the question answers itself: Better logic than nothing, and better nothing than God. It is a response that serves moral as well as physical thought. Philosophers such as Simon Blackburn, who believe that it is their special responsibility to decline theological appeals, also find themselves forced to choose between logic and nothing.

It is a choice that offers philosophers and physicists little room in which to maneuver. All attempts to see the laws of nature as statements that are true in virtue of their form have been unavailing. The laws of nature, as Isaac Newton foresaw, are not laws of logic, nor are they *like* the laws of logic. Physicists since Einstein have tried to see in the laws of nature a formal structure that would allow them to say to themselves, "Ah, that is why they are true," and they have failed. Before determining that he would welcome in the form of the Landscape and the Anthropic Principle ideas that previously he was prepared to reject, Steven Weinberg argued that when at last we are face-to-face with the final theory, we shall discover that it is *unique*. It is what it is. It cannot be changed. And it is precisely the fact that it cannot be changed that offers the soul surcease from anxiety. In the end, this idea does not serve the cause. If it is *impossible* to change the structure of a final theory, then uniqueness is simply a coded concept, one standing for necessity itself. And if it is not impossible, the claim that the final laws of nature are unique comes to little more than this: They are what they are, and who on earth knows why?

While better logic than nothing is still on the menu, it is no longer on the table. There remains better nothing than God as the living preference among physicists and moral philosophers. It is a remarkably serviceable philosophy. In moral thought, nothing comes to moral relativism; and philosophers who can see no reason whatsoever that they should accept any very onerous moral constraints have found themselves gratified to discover that there are no such constraints they need accept. The Landscape and the Anthropic Principle represent the ascendance of moral relativism in physical thought. They work to cancel the suggestion that the universe—our own, the one we inhabit—is any kind of put-up job. This is their emotional content, the place where they serve prejudice. These ideas have an important role to play in the economy of the sciences, and for this reason, they have been welcomed by the community of scientific atheists with something akin to a cool murmur of relief. They have, for example, worked entirely to

Richard Dawkins's satisfaction. He believes them superior to the obvious theological alternatives on the grounds that it is better to have many worlds than one God.

But before his enthusiasm is dismissed as obviously contrived, it should be remembered that just these principles *have* led to a startling physical prediction. Using the ideas of the Landscape and the Anthropic Principle, Steven Weinberg predicted that the cosmological constant, as it is observed, should have a small, positive value. In this he was correct. This is very remarkable and it suggests that just possibly these ideas have a depth somewhat at odds with their apparently frivolous character.

I do not know. It does not hurt to say so.

But one possibility in thought should certainly encourage another. If nothing proves unavailing, will physicists accept the inexorable logic of the disjunction *God or nothing*?

Writing with what I think is characteristic honesty, Leonard Susskind has this to say:

> If, for some unforeseen reason, the landscape turns out to be inconsistent—maybe for mathematical reasons, or because it disagrees with observation—I am pretty sure that physicists will go on searching for natural explanations of the world. But I have to say that if that happens, as things stand now we will be in a very awkward position. Without any explanation of nature's fine-tunings we will be hard pressed to answer the ID [intelligent design] critics. One might argue that the hope that a mathematically unique solution will emerge is as faith-based as ID.

This remark has an unintended daring. It gives a good deal of ground away. It is generous. And it suggests oddly enough that a conflict in thought that scientists have almost universally dismissed retains a strange, disturbing vitality nonetheless. Do not be misled by phrases such as "faith-based as ID." It is the word *awkward* that counts. If the double ideas of the Landscape and the Anthropic Principle do not suffice to answer the question why we live in a universe that seems perfectly *designed* for human life, a great many men and women will conclude that it *is* perfectly designed for human life, and they will draw the appropriate consequences from this conjecture.

What is awkward is just that at a moment when the community of scientists had hoped that they had put all that behind them so as to enjoy a universe that was safe, sane, secular, and sanitized, somehow the thing they had been so long avoiding has managed to clamber back into contention as a living possibility in thought.

This is *very* awkward.

7

A Curious Proof That God Does Not *Exist*

A N ARGUMENT *for* God's existence is a commonplace; an argument *against* his existence is an event. Just such an argument comprises the centerpiece of Richard Dawkins's *The God Delusion*. It is an argument to which he attaches the utmost importance: In lectures and talks given since its publication, he has suggested that it now looms very large in the troubled imagination of religious believers. In this he is mistaken. His argument is nonetheless important in another respect. It is an object lesson.

Dawkins summarizes his views in a series of six very general propositions, of which only the first three are directly relevant to his concerns—or mine:

The first affirms that the universe is improbable.

The second acknowledges the temptation to explain the appearance of the universe by an appeal to a designer.

And the third rejects the temptation on the grounds that "the designer hypothesis immediately raises the larger problem of who designed the designer." A variant of this argument has been known for a very long time.

"I venture to ask," the Chinese sage Kuo Hsiang ventured to ask in the third century A.D., "whether the Creator is or is not. If He is not, how can He create things? And if He is, then (being one of those things), He is incapable (without self-creation) of creating the mass of bodily forms."

This argument is exquisite because it is short.

Persuaded that God does not exist, Richard Dawkins might have quoted Kuo Hsiang and left matters there.

As it happens, Dawkins presents *his* argument in the first two pages of chapter 4 of his book and summarizes it in the chapter's last two pages. The material in between—some forty pages—is given over to the "consciousness raising" that contemplation of natural selection is said to evoke. In all this, Dawkins has failed only to explain his reasoning, and I am left with the considerable inconvenience of establishing his argument before rejecting it.

THE DEAD ZONE

As a public figure and so a character in debate, Richard Dawkins may be found in the dead zone marking the intersection of a child's question—"Who made God?"—and what the classicist R. R. Bolgar called "the peculiar debris of an abandoned and virtually forgotten science." Although discussing rhetoric, Bolgar could well have been describing theology. The zone is dead because the questions it encourages are unanswerable. This hardly means that they are insignificant. Childish questions have their point, and in the case of God's existence, their point is to place in doubt some of the intellectual maneuvers by which His existence is affirmed.

Doubt is very much a matter of temperament. It is rarely encouraged (or displaced) by argument. For certain temperaments, the existence of the universe is a mystery, one that gnaws irritably at the soul. Why *is* the damn thing there? The thought that it is there for no good reason is said by some to spoil their enjoyment of life. In time taken from writing, hunting, estate management, and fornication, Leo Tolstoy very often expressed sentiments such as this. Like Levin in *Anna Karenina,* he was widely regarded as a pest for doing so. On the other hand, a great many men and women take the universe in stride, and if they are disposed to ask why it is there, they are easily pleased with the answer that the physicist (and Nobel laureate) Frank Wilczek insouciantly offered: "The universe," he wrote, "appears to be just one of those things." A willingness to let the matter rest in this way is a characteristic of individuals that William James described as "healthy-minded"—another way of describing them as thick.

Of course, if physicists can believe that the universe is just one of those things, then believers can affirm that God is just one of those things as well.

To the question of why believers should not stop with the universe, there is only the counterquestion of why physicists should not proceed further to God.

I mention these points to stress what should be obvious: Questions arising in the dead zone *are* a matter of temperament. A religious instinct is universal: It arises in every human being—hence the popular observation that there are no atheists in foxholes. But whether an instinct is allowed to progress toward frank affirmation, or whether it is denied and then discarded—these are not issues that answer to any obvious claims of argument.

This is one reason the dead zone is dead.

❧

If God did not create the world, then what is His use? And if He did, then what is His explanation? A child's question has given way to an adult's dilemma. A God too indisposed to do the work of creation is fated to drift into irrelevance, if only because His demand for adoration would be considerably out of line with His record of accomplishment. But if God *did* create the world, the problem that God is designed to solve reappears as a problem about God Himself.

It is this destructive dilemma that Dawkins calls the Ultimate Boeing 747 gambit. The appeal to a Boeing 747 is meant to evoke a lighthearted quip attributed to the astrophysicist

Fred Hoyle. The spontaneous emergence of life on earth, Hoyle observed, is about as likely as a tornado sweeping through a junkyard and assembling a Boeing 747 out of the debris. Although an atheist, Hoyle was skeptical about Darwin's theory of evolution, and Dawkins passionate in its defense. Since the junkyard expresses with rare economy precisely the odds favoring the spontaneous appearance of life—they are remarkably prohibitive on virtually every calculation—it has been an irritation to Dawkins ever since it made its appearance. With their consciousness unraised, a great many people have evidently concluded that when it comes to the origins of life, the junkyard is all that Darwin offered.

But, Dawkins affirms, if a tornado cannnot do the job of creating life, then God cannot do the job of creating the universe. The tornado is inadequate because life is improbable, and God is inadequate for the same reason. This counterstroke has persuaded Dawkins that he has initiated an intellectual maneuver judo-like in its purity of effect and devastating in its consequences. The Ultimate 747 gambit, Dawkins writes, "comes close to *proving* that God does *not* exist" (italics added). Fred Hoyle's death before he could appreciate the extent of his discomfiture Dawkins no doubt regards as a display of peevish irresponsibility.

Although Dawkins writes with quiet confidence about what he intends to do, which is to give the Deity a thrashing, and then, having thrashed him, writes again about what he has done, what he is doing is rather less clear.

At times, Dawkins asserts that God is an irrelevance because He has been assigned the task of constructing a universe that is improbable. If the universe is improbable, "it is obviously no solution to postulate something even more improbable." Why an improbable universe demands an improbable God, Dawkins does not say and I do not know.

There are other passages in *The God Delusion* of more analytic refinement. In these, Dawkins extends a convivial pseudopod toward the concepts of complexity and information. Under the influence of these concepts, Dawkins often writes that unless God is Himself complex, He is bound to be inadequate to account for the complexity of the universe. The very same observation he sometimes makes in terms of information. If Dawkins is casual about these concepts to the point of slackness, it is because he believes that whether his argument is expressed in terms of information *or* complexity, God will emerge with His irrelevance undiminished.

The 747 gambit, although hardly a model of fastidiousness, conveys a beefy impression of authority, so much so that scientists who never once thought seriously about issues of religion at once wondered why they did not think of it themselves. Having not thought of it at all, they often appear to have thought of it after all. Publishing his thoughts in *Gene* (of all places!), the distinguished molecular geneticist Emile Zuckerkandl has argued that the Deity, if He exists, would represent "something like a pathology of the state of being." I had very much hoped that after beginning with pathology, Zuckerkandl would continue to some form of exciting degeneracy, but it was not to be; what Zuckerkandl in the end does offer is homegrown but homeopathic, a dilute solution of the 747 gambit. His target is very much theories of intelligent design. Designating the intelligent designer as the Higher Intelligence, he writes that "if complexity is a problem for naturalistic explanations, the Higher Intelligence itself is first to have to face this problem. Intelligent Design thus does not solve any problem posed by complexity; it only transposes the origins of complexity from the observable to an unobservable world and makes these origins inaccessible to inquiry."

These are words that display a somewhat Teutonic sternness in attitude. Less demanding critics might observe that shoveling problems backward until they are out of sight is not only the tactic of common sense but the only tactic in common use. When scientists appeal to various unobservable entities—universal forces, grand symmetries, twice-differential functions as in mechanics, Calabi-Yau manifolds, ionic bonds, or quantum fields—the shovel is in plain sight, but what has been shoveled is nowhere to be seen. Why physicists should enjoy inferential advantages denied theologians, Zuckerkandl does not say.

The difficulty with these arguments—they form a genre—is that they endeavor to reconcile two incompatible tendencies in order to force a dilemma. On the one hand, there is the claim that the universe is improbable; on the other, the claim that God made the universe. Considered jointly, these claims form an unnatural union. Probabilities belong to the world in

which things happen because they might, creation to the world
in which things happen because they must. We explain creation by appealing to creators, whether deities or the inflexible
laws of nature. We explain what is chancy by appealing to
chance. We cannot do both. If God did make the world, it is
not improbable. If it is improbable, then God did not make it.
The best we could say is that God made a world that *would*
be improbable *had* it been produced by chance.

But it wasn't, and so He didn't.

This is a discouraging first step in an argument said to
come close to proving that God does not exist.

AN UNLIKELY DEITY

Let us say that Dawkins is quite right. God is improbable.
The proposition is on the table. It is up for grabs.

What then follows?

Oddly enough, almost nothing. I say "oddly enough" because the thesis that God's existence is improbable is the cor-
nerstone of Dawkins's argument and so an easement to his
atheism. But atheists have traditionally been concerned not
to push too vigorously on their own suspicions by presenting
them in a way that owes too much to probability. The stuff—
probability, I mean—is notoriously unstable.

The inference that Dawkins proposes to champion has as
its premise the claim that God is improbable; its conclusion is
that likely God does not exist. The inferential bridge invoked
by the 747 gambit, if it goes anywhere at all, goes from what

God is (He is unlikely) to whether He exists (it would appear not). Inferences of this sort are typically not deductive: they do not impart certainty to their conclusions. A deductive inference carries conviction straight on down. *All men are mortal.* Premise One. *Socrates is a man.* Premise Two. And the conclusion. *Socrates is mortal.* Given the premises, the conclusion is incontestable.

The attempt to shoehorn inferences hinging on likelihood into deductive form ends in disaster. An example? *Judging by how Emile writes, it is likely that his native language is German.* Premise One. *Judging by where Emile lives, it is likely that his native language is English.* Premise Two. Whatever conclusions may be drawn from the circumstances of Emile's prose and his residence, they are not deductive. Otherwise, the result would be a contradiction, Emile afflicted with at least two native languages, and possibly more. It may well be—I am supposing— that certain considerations make God's existence unlikely. Other considerations might make his existence likely. If both considerations are deductively controlling, the result is a form of logical chaos. The inference Dawkins champions cannot *prove* anything about God's existence, and if it cannot prove anything about God's existence, it cannot come *close* to proving anything either.

There is next the explosion that results when improbability and existence are foolishly mingled. In this regard it is curious that having declared God's existence unlikely in virtue of His improbability, Dawkins never once considers that by parity of reasoning he could well have concluded that the existence of the universe is unlikely in virtue of *its* improbability. Unlikely is unlikely, as logicians say, never adding, of course, that if the universe is unlikely there is the slightest reason to suppose that it does not exist. Nonetheless, the assumption that the universe is improbable is the gravamen of the 747 gambit. It is indispensable.

The fact is that unlikely events do occur. They simply do not occur often. It is as difficult, the Bible recounts, for a rich man to enter the kingdom of heaven as it is for him to pass through the eye of a needle; and if it is that difficult, I suppose it is that improbable. And yet some rich men manage. I might by this reasoning at least *anticipate* meeting King Farouk in the hereafter, where, having squeezed himself through the eye of that needle, he may be found enjoying the celestial gaming tables. It *could* happen. As quantum theorists never tire of reminding us, what could happen sooner or later will happen.

What holds for Farouk might well hold for the Deity.

Having brought the universe into existence, *He* might simply be improbable. It is just one of those things for which we have no *further* explanation.

This is not a conclusion that the soul on fire to know God will necessarily find discouraging. But if we have no further explanation for the existence of the Deity, we nonetheless do by this means have an explanation for the existence of the *universe*. It is this: *In the beginning God created the heaven and earth.*

❦

Arguments trading on probability, I have suggested, are unstable. Like certain women, they go off at the worst possible moments. The theory of probability is in the business of assigning numbers to events. The theory assumes explicitly what everyone ordinarily takes for granted, and that is that if events are assigned probabilities, they are determined by means of a random process. An improbable God must thus be improbable in virtue of the process that controls his probability. Just *which* random process is designed to yield the Deity as a possible outcome? It is by no means easy to say, which is one reason, I suppose, that on this subject, Dawkins says nothing at all.

Whatever the process, the probabilities it reveals depend on the way they are described. Events that are improbable over the short term become probable and even certain over the long term, as when, typing randomly in solitude, a single monkey— a great ape in virtue of accomplishments to come—re-creates Shakespeare's *Hamlet,* with every comma in place and variant spellings noted. No one expects this prodigy to finish up *quickly,* of course, which is another way of saying that it all depends.

An improbable God, denied access to Being over the short term, may find himself clambering into existence over a term that is long. But having failed to control the circumstances by which God's probability is assigned, Dawkins has also neglected to mention how long those circumstances have been in operation. We are thus free to imagine some infernal cosmic experiment involving dice whose clatter is intended to evoke deities, with each round failing to elicit the Deity, until God *finally* appears, brimming with enthusiasm and ready to create the universe. So far as God is concerned, after all, He has all the time in the world.

It all depends, of course.

EXPLANATIONS WITHOUT END

When expressed as Dawkins expresses it, the Ultimate 747 gambit explodes and then gutters out inconclusively. But often arguments of this sort carry with them a shadow, one prepared in a pinch to take over from the main character. In the case of the 747 gambit, the shadow is given over to meditations concerning the structure of rational explanations. It is an important topic, and one that Dawkins is prepared to cover in the cloak of his carelessness.

A single power assumption is at work throughout: Unlikely events require an explanation. The power assumption trails in its wake two additional assumptions. The first is that old standby: The universe is unlikely. And the second has long stood by: *If* God created the universe, He must be more unlikely than the universe He created. From the power assumption and its sidekicks, an infinite regress very quickly arises, one in which God requires an explanation, which in turn triggers the demand for yet another explanation, and so another God.

It follows that if God created the universe, there are Gods stacked up behind him, each one creating the God below.

Either we must give up on him (formerly Him) or we need to get to the God who really gets things going, and since there seems by this argument to be infinitely many of them, each presumably more powerful, and certainly more intimidating, than His Subordinates, this is an inquiry guaranteed either to fail or to lead to a revival of an especially vigorous form of polytheism.

Imagine addressing prayers to our *fathers* which art in heaven, as in one of those ghastly children's books in which Heather has three mothers and Jamal a dozen fathers.

Depressing, no?

The demand that explanations mention only events no more improbable than the events they explain is in any case intolerably abstemious. "How often have I said to you," Sherlock Holmes observed to Watson, "that when you have eliminated the impossible, whatever remains, *however improbable*, must be the truth?"

But if we explain an event by an appeal to an improbable event, it does not follow that we are bound to keep clambering up the ladder of an infinite regress. When in *The Perfect Storm* Sebastian Junger described a freak storm off the coast of Nova Scotia, he was explaining shipwreck at sea by means of a very rare confluence of meteorological factors. Such explanations are common throughout the sciences and they are common in ordinary life. About such rare events, all that we can say is that sometimes they happen. We do not need to say more. What more could we say? It may well be that God is improbable and that is the end of it. When Christian believers give thanks for the miracle of Christ, they mean by a miracle a *miracle*.

THE INADVERTENT THEIST

Although Richard Dawkins has nothing but contempt for theology, often glorying in his impressive ignorance, with his argument he finds himself occupying an unexpected position of prominence amid "the peculiar debris of an abandoned and virtually forgotten science."

In addition to suffering the infirmity of improbability, the God whose existence Dawkins is prepared to challenge seems curiously a diminished figure. He has gotten the job of creation done. His time thereafter has been spent imposing onerous sexual constraints upon the Jewish people and when absolutely required undertaking a miracle or two. For the moment, He seems to have vacated the universe with a smashing headache. On his previous appearances, He seemed very much like a lumbering robot. One might almost expect to hear the lingering echos of divine clanking. Above all, He is very much a *contingent* deity. If He is here today, He may be gone tomorrow. If His existence were guaranteed, the argument that Dawkins has advanced would fail before it started instead of starting before it failed.

And yet these are considerations long familiar in the history of theological thought. They form the heart and soul of Aquinas's second cosmological argument, and if Aquinas gives them only a few words, that is because he requires only a few words to say what needs to be said. Any conception of a contingent deity, Aquinas argues, is doomed to fail, and it is doomed to fail precisely because whatever He might do to explain the existence of the universe, His existence would again require an explanation. "Therefore, not all beings are merely possible, but there must exist something the existence of which is necessary."

The conclusion that a religious believer will take from Dawkins's argument is either that God is improbable or that He is *necessary*.

What Dawkins has established serves chiefly as a reminder: Explanations come to an end, and because we are human, they must come to an end before they have satisfied every one of our emotional needs. But scientific atheists should at least be open to the possibility that scientific explanations by their very nature come to an end well before they have done all the work that an explanation can do. If they have not read Aquinas's *Summa Theologica*, physicists have nonetheless heard its music. They have hoped to discover laws of some final physical theory so powerful that they will explain the property of matter in all its modes. "The most extreme hope for science," Steven Weinberg has written, "is that we will be able to trace the explanation of all natural phenomena to final laws and historical accidents."

This is the most extreme hope for science for those, like Frank Wilczek, inclined to say at some point that that's just how things are. For others, intellectual comfort is less easily purchased. "We feel," Wittgenstein wrote, "that even when *all possible* scientific questions have been answered, the problems of life remain completely untouched." Those who do feel this way will see, following Aquinas, that the only inference calculated to overcome the way things are is one directed toward the way things must be.

Perhaps in the end this will prove to be a matter of mathematics. MIT physicist Max Tegmark has argued that this is so. The physicist Edward Witten and the mathematician Alain Connes have both written suggestively about the origins of creation in some inexplicably austere and remote mathematical structure, one so powerful that from it space and time themselves may be derived.

With these ambitious speculations on the table, it is worthwhile to recall that in locating the origins of creation in some fundamental abstract structures, mathematicians are assigning to them a degree of agency that until now they do not seem to have possessed.

⟨⟩

There remains a final point. What a man rejects as distasteful must always be measured against what he is prepared eagerly to swallow. What Richard Dawkins *is* prepared to swallow is the Landscape and the Anthropic Principle. The Landscape does not, of course, answer the question what caused the Landscape to exist. How could it? And if nothing caused the Landscape, it does not answer the question why it should be there at all.

But having swallowed the Landscape with such inimitable gusto, Dawkins is surely obliged to explain just why he scruples at the Deity. After all, the theologian need only appeal to a single God lording over it all and a single universe—our own. Dawkins must appeal to infinitely many universes crammed into creation, with laws of nature wriggling indiscreetly and fundamental physical parameters changing as one travels from one corner of the cosmos to the next, the whole entire gargantuan structure scientifically unobservable and devoid of any connection to experience.

This is a point that Dawkins endeavors to meet, but with markedly insufficient success. "The key difference between the radically extravagant God hypothesis," he writes, "and the apparently extravagant multiverse hypothesis, is one of statistical improbability."

It is? I had no idea, the more so since Dawkins's very next sentence would seem to undercut the sentence he has just written. "The multiverse, for all that it is extravagant, is simple," because each of its constituent universes "is simple in its fundamental laws."

If this is true for each of those constituent universes, then it is true for our universe as well. And if our universe is simple in its fundamental laws, what on earth is the relevance of Dawkins's argument?

Simple things, simple explanations, simple laws, a simple God.

Bon appétit.

CHAPTER

8

Our Inner Ape, a Darling, and the Human Mind

THE IDEA that human beings have been endowed with powers and properties not found elsewhere in the animal kingdom—or the universe, so far as we can tell—arises from a simple imperative: *Just look around.* It is an imperative that survives the invitation fraternally to consider the great apes. The apes are, after all, behind the bars of their cages and we are not. Eager for the experiments to begin, they are impatient for their food to be served. They seem impatient for little else. After years of punishing trials, a few of them have been taught the rudiments of various primitive symbol systems. Having been given the gift of language, they have nothing to say. When two simian prodigies meet, they fling their signs at one another.

155

More is expected, but more is rarely forthcoming. Experiments conducted by Dorothy Cheney and Robert Seyfarth—and they are exquisite—indicate that like other mammals, baboons have a rich inner world, something that only the intellectual shambles of behavioral psychology could ever have placed in doubt. Simian social structures are often intricate. Chimpanzees, bonobos, and gorillas reason; they form plans; they have preferences; they are cunning; they have passions and desires; and they suffer. The same is true of cats, I might add. In much of this, we see ourselves. But beyond what we have in common with the apes, we have nothing in common, and while the similarities are interesting, the differences are profound.

If human beings are as human beings think they are, then religious ideas about *what* they are gain purchase. These ideas are ancient. They have arisen spontaneously in every culture. They have seemed to men and women the obvious conclusions to be drawn from just looking around. An enormous amount of intellectual effort has accordingly been invested in persuading men and women not to look around. "The idea that human minds are the product of evolution is 'unassailable fact.'" Thus *Nature* in an editorial. Should anyone have missed the point, *Nature* made it again: "With all deference to the sensibilities of religious people, the idea that man was created in the image of God can surely be put aside."

Those not willing to put such sentiments aside, the scientific community has concluded, are afflicted by a form of intellectual ingratitude.

It is remarkable how widespread ingratitude really is.

ALFRED WALLACE: A DISSENT

Together with Charles Darwin, Alfred Wallace created the modern theory of evolution. He has been unjustly neglected by history, perhaps because shortly after conceiving his theory, he came to doubt its provenance. Darwin, too, had his doubts. No one reading *On the Origin of Species* could miss the note of moral anxiety. But Darwin's doubts arose because, considering its consequences, he feared his theory might be true; with Wallace, it was the other way around. Considering its consequences, he suspected his theory might be false.

In an interesting essay published in 1869 and entitled "Sir Charles Lyell on Geological Climates and the Origin of Species," Wallace outlined his sense that evolution was inadequate to explain certain obvious features of the human race. The essay is of great importance. It marks a falling-away in faith on the part of a sensitive biologist previously devoted to ideas he had himself introduced. Certain of our "physical characteristics," he observed, "are not explicable on the theory of variation and survival of the fittest." These include the human brain, the organs of speech and articulation, the human hand, and the external human form, with its upright posture and bipedal gait. It is only human beings who can rotate their thumb and ring finger in what is called ulnar opposition in order to achieve a grip, a grasp, and a degree of torque denied any of the great apes. No other item on Wallace's list has been ticked off against real understanding in evolutionary thought. What remains is fantasy of the sort in which the bipedal gait is assigned to an unrecoverable ancestor wishing to peer (or pee) over tall savannah grasses.

The argument that Wallace made with respect to the human body he made again with respect to the human mind. There it gathers force. Do we understand why alone among the animals, human beings have acquired language? Or a refined and delicate moral system, or art, architecture, music, dance, or mathematics? This is a severely abbreviated list. The body of Western literature and philosophy is an extended commentary on human nature, and over the course of more than four thousand years, it has not exhausted its mysteries. "You could not discover the limits of soul," Heraclitus wrote, "not even if you traveled down every road. Such is the depth of its form."

Yet there is no evident distinction, Wallace observed, between the mental powers of the most primitive human being and the most advanced. Raised in England instead of the Ecuadorian Amazon, a native child of the head-hunting Jívaro, destined otherwise for a life spent loping through the jungle, would learn to speak perfect English, and would upon graduation from Oxford or Cambridge have the double advantage of a modern intellectual worldview and a commercially valuable ethnic heritage. He might become a mathematician, he would understand the prevailing moral and social codes perfectly, and for all anyone knows (or could tell), he might find himself a BBC commentator, explaining lucidly the cultural significance of head-hunting and arguing for its protection.

From this it follows, Wallace argued, that characteristic human abilities must be latent in primitive man, existing somehow as an unopened gift, the entryway to a world that primitive man does not possess and would not recognize.

But the idea that a biological species might possess latent powers makes no sense in Darwinian terms. It suggests the forbidden doctrine that evolutionary advantages were front-loaded far away and long ago; it is in conflict with the Darwinian principle that useless genes are subject to negative selection pressure and must therefore find themselves draining away into the sands of time.

Wallace identified a frank conflict between his own theory and what seemed to him obvious facts about the solidity and unchangeability of human nature.

The conflict persists; it has not been resolved.

WHAT NO ONE DOUBTS

No one doubts that human beings now alive are connected to human beings who lived thousands of years ago. To look at Paleolithic cave drawings is to understand that the graphic arts have not in twelve thousand years changed radically. And no one doubts that human beings are connected to the rest of the animal kingdom.

It is rather more difficult to take what no one doubts and fashion it into an effective defense of the thesis that human beings are *nothing but* the living record of an extended evolutionary process. That requires a disciplined commitment to a point of view that owes nothing to the sciences, however loosely construed, and astonishingly little to the evidence.

It is for this reason—no science, little evidence—that the kinship between human beings and the apes has been promoted in contemporary culture as a moral virtue as well as a zoological fact. It functions as a hedge against religious belief, and so it is eagerly advanced. The affirmation that human beings are fundamentally unlike the apes is widely considered a defect of prejudice or a celebration of trivialities. "Chimps and gorillas have long been the battleground of our search of uniqueness," Stephen Jay Gould remarked, "for if we could establish an unambiguous distinction—of kind rather than degree—between ourselves and our closest relatives, we might gain the justification long sought for our cosmic arrogance."

Following Stephen Jay Gould, whose "cool authentic voice" he finds irresistible, Christopher Hitchens endorses the Master in declaring against cosmic arrogance. I may well be its last supporter, all things considered. "If the numberless evolutions from the Cambrian period could be recorded and 'rewound,'" Hitchens writes, "and the tape played again, he [Gould] established there was no certainty it would come out the same way." Having no access to the tape of life, Gould established nothing of the sort, of course; I am recounting the story line purely *pour le sport*. And what sport it is, involving as it does only the celebration of an obvious tautology. Had an early vertebrate named Pakaia not survived, its survivors, Hitchens reports in amazement, would not have survived. No deflation of arrogance could be more rigorous. Or less interesting. I would find Hitchens's thoughts even more gratifying than I do had he not enlarged them to encompass nonlinear dynamics and Heisenberg's uncertainty principle, subjects that in his ineptitude he waves like a majestic frond.

When it comes to the apes, the argument is so uncertain that it must be made with the assurance that arises from the affirmation of an absurdity. In writing about "our inner ape," Frans de Waal is thus concerned to demonstrate "how much apes resemble us and how much we resemble them."

How much, then, do we resemble them, or they us? No, really? The correct answer, of course, is that although we resemble apes in some ways, we are nonetheless quite different, and we are different in ways that are of great biological and moral importance. If this is the correct answer, it is not the one de Waal is proposed to endorse. "If an extraterrestrial were to visit earth," de Waal writes, "he would have a hard time seeing most of the differences we treasure between ourselves and the apes."

I suppose that if a fish were thoughtfully to consider the matter, she might have a hard time determining the differences we treasure between Al Gore and a sperm whale. Both of them are large and one of them is streamlined. This is, perhaps, one reason fish are not more often consulted on important matters of taxonomy. Or anything else.

Wishing a more detailed (but no more obvious) demonstration, both the fish and de Waal's extraterrestrial would profit from reading a fundamental paper on the subject. Writing in *Science* in 1975, M.-C. King and A. C. Wilson provided for the first time an estimate of the degree of similarity between the human and the chimpanzee genome. Far more than was thought possible at the time, King and Wilson claimed, human beings and chimpanzees share the greater part of their respective genomes.

Whence the conclusion that if our genomes match up so nicely, we must *be* apes.

In the second section of their paper, King and Wilson describe honestly the deficiencies of this idea. Human beings and the apes, they observe,

> differ far more than sibling species in anatomy and way of life. Although humans and chimpanzees are rather similar in the structure of thorax and arms, they differ substantially not only in brain size but also in the anatomy of the pelvis, foot, and jaws, as well as in relative lengths of limbs and digits. Humans and chimpanzees also differ significantly in many other anatomical respects, to the extent that nearly every bone in the body of a chimpanzee is readily distinguishable in shape and size from its human counterpart. Associated with these anatomical differences there are, of course, major differences in posture, mode of locomotion, methods of

procuring food, and means of communication. Because
of these major differences in anatomy and way of life,
biologists place the two species not just in separate gen-
era but in separate families.

There is nothing in this that was not evident to Alfred
Wallace. Or to any student of comparative anatomy. King and
Wilson went on to suggest that the morphological and behavioral differences between humans and the apes, if they were
not due to variations between their genomes, must be due to
variations in their genomic regulatory systems. These are the
systems that control the activities of the genes by telling various genes when to sound off and when to shut up. They are of
an astonishing and poorly understood complexity, if only because they themselves require regulation. Higher-order regulation in turn involves higher-order codes beyond the genetic
code. Codes then require their own regulation. Even the simplest cell involves an intricate, never-ending cascade of control
and coordination of a sort never seen in the physical world. It
is entirely safe to assign the differences between human beings
and the apes to their regulatory systems. Nothing is known
about their evolutionary emergence and we cannot describe
them with any clarity.

Whatever the source of the human distinction in nature,
its existence is obvious, and when it is carelessly denied, the
result is a very characteristic form of inanity.

Thus Jonathan Gottshall recounts his experiences in reading Homer's *Iliad* while under the influence of the thesis, as he

puts it, that "people are apes." It is a thesis that he attributes to Desmond Morris's *The Naked Ape*. "But this time around," Gottshall explains, "I also experienced the *Iliad* as a drama of naked apes—strutting, preening, fighting and bellowing their power in fierce competition for social dominance, beautiful women and material resources." Social dominance and material resources are, in fact, not quite to the point. "Intense competition between great apes, as described both by Homer and by primatologists, frequently boils down to precisely the same thing: access to females."

The governing words in this quotation are "boils down," and as in so many such analyses, what is essential is not what has been distilled but what has evaporated.

That is, everything of interest in the *Iliad*.

At the height of the battle of Stalingrad, a young lieutenant with the 24th German Panzer Division wrote in his diary that Stalingrad "by day, is an enormous cloud of burning blinding smoke; it is a vast furnace lit by the reflection of the flames. And when night comes, one of those scorching, howling, bleeding nights, the dogs plunge into the Volga and swim desperately to gain the other bank. The nights of Stalingrad are a terror for them. Animals feel this hell, the hardest stones cannot endure it; *only men endure*" (italics added).

Would anyone reading these words imagine that man's endurance is remotely like the zestful competition of apes eager to copulate and vexed when they cannot?

This suggests an obvious counsel of humility. It is one that may profitably be directed toward biologists overmuch worried about cosmic arrogance. Before putting aside so carelessly "the idea that man was created in the image of God," first consider the ideas you propose to champion in its place. If they are no good, why champion them?

And they are no good. So why champion them?

THE DARLING

Edward Wilson published *Sociobiology* and Richard Dawkins *The Selfish Gene* during the 1970s. Since then, evolutionary psychology has become a contemporary darling. The story that it advances is one that takes place entirely within the human species. No apes need apply, for none are wanted.

The essentials are simple and they have the simple-minded structure of a fairy tale—indeed, the philosopher David Stove entitled his attack on evolutionary psychology *Darwinian Fairytales*. The significant features of human psychology first arose during the late Paleolithic era—the so-called Era of Evolutionary Adaptation. For reasons that no one has properly specified, it was then that human beings devised their responsive strategies to the contingencies of life—getting food, getting by, and getting laid. These strategies have persisted to the present day. They are at the core of the modern human personality. We are what we were. There followed the long Era in Which Nothing Happened, the modern human mind retaining in its structure and programs the mark of the time that human beings spent in the savannah or on the forest floor, hunting, gathering, and reproducing with Darwinian gusto.

If their content is negligible, the influence of these stories is immense. Commenting on negative advertising in political campaigns, Kathleen Hall Jamieson, the director of the Annenberg Public Policy Center at the University of Pennsylvania, remarked that "there appears to be something hardwired into humans that gives special attention to negative information." There followed what is by now a characteristic note: "I think it's evolutionary biology." The fact that there is nothing hardwired about human beings, because they are not wired at all, is passed over as incidental. The metaphor has taken on a life all its own, and now that it is living, it has grown great.

Having provided an explanation of negative campaign advertisements, evolutionary biology also explains war and male aggression, the human sensitivity to beauty, gossip, a preference for suburban landscapes, love, altruism, marriage, jealousy, adultery, road rage, religious belief, fear of snakes, disgust, night sweats, infanticide, and the fact that parents are often fond of their children. The idea that human behavior is "the product of evolution," as the *Washington Post* puts the matter, is now more than a theory, it is a popular conviction. The conviction is so popular that it may even be impudently opposed to the regnant conventions of political correctness. The result is an inspiring clash of clichés. In an article published in *Psychology Today*, Alan S. Miller and Satoshi Kanazawa describe what they consider ten politically incorrect truths about human nature. They regard these truths as *discoveries,* and in recounting the first item on their list report with an explorer's sense of satisfied astonishment—*Would you look at that?*—that "men like blonde bombshells." If this is a truth about human nature, it has not been especially well hidden.

Nor does it seem to cry out for any explanation beyond the obvious. Men like blonde bombshells because they *are* blonde bombshells. I would even go further. Men seem to like bombshells no matter the color of their hair.

What more is really needed?

For Miller and Kanazawa, the demands of science are more considerable. What they require is an explanation *beyond* the obvious. *Aussitôt dit, aussitôt fait,* as the French say. No sooner said than done. Our ancestors millions of years ago, they assert, were evidently concerned to discover fine, healthy women, and lacking suitable gynecological skills, necessity compelled them to pay attention to their secondary sexual characteristics. Whence the popularity of blonde bombshells. The subject has been the focus of research extending over decades, psychologists investigating the extent to which blonde bombshells *are* bombshells with never-flagging zeal and, in some cases, even conducting their research in various strip clubs the better to establish that various bombshells really are blonde.

Unneeded as an irrelevance, these ideas are implausible as an explanation. If sexual preferences are rooted in the late Paleolithic era, men worldwide should now be looking for stout muscular women with broad backs, sturdy legs, a high threshold to pain, and a welcome eagerness to resume foraging directly after parturition. It has not been widely documented that they do.

Our ancestors are in any case unavailable. Claims made on their behalf are unverifiable. The underlying tissue that connects the late Paleolithic and the modern era is the gene pool. Changes to that pool reflect a dynamic process in which genes undergo change, duplicate themselves, surge into the future or shuffle off, and by means of all the contingencies of life serve in each generation the purpose of creating yet another generation. It is precisely these initial conditions that popular accounts of human evolution cannot supply. We can say of those hunters and gatherers only that they hunted and they gathered, and we can say this only because it seems obvious that there was nothing else for them to do. The gene pool that they embodied cannot be recovered.

The largest story told by evolutionary psychology is therefore anecdotal. It has no scientific value.

We might as well be honest with one another. It has no value whatsoever.

THE HUMAN MIND

It was just yesterday that Freud depicted the human mind as a fabulous haunted house. The image has had an enduring value, if only because on some level we are all haunted by things we cannot name and do not recognize. The analytical deficiencies of Freudian theory are nonetheless considerable, for if the Freudian house was haunted, Freud was unable to say *who* haunted it. Items such as the id, the ego, and the super-ego functioned as characters in Freud's system. They had needs, they made their demands known, they were artful in concealment; and these are among the attributes of the human mind for which an explanation was originally needed.

The haunted house has given way in our time to the digital computer. The argument proceeds in steps. The first involves the dismissal of the mind as a separate ontological category. Mind *and* brain, as Descartes supposed? This is Cartesian dualism and widely rejected by philosophers. "Every aspect of thought and emotion is," Steven Pinker has argued in *How the Mind Works*, "rooted in brain structure and function." This tells us where the mind is by telling us where it has gone to. It has been swallowed up by the brain. If one needs directions, it suffices to tap significantly at one's skull:

Tap One: The mind is the brain.

How a craving for raspberry Jell-O might be located within the human brain, Pinker does not say. Perhaps it involves neurons devoted to gelatin? I am asking in a spirit of honest inquiry.

With the mind removed, it remains for Pinker to explain how the brain undertakes so many activities formerly undertaken by the mind. As it happens, this is not a problem either. It is the brain's capacity to process information, Pinker believes, that "allows human beings to see, think, feel, choose and act." The digital computer is precisely a device designed to process information. Whereupon we discover how the mind works. The same significant tap may be invoked for a second time.

Tap Two: The brain is a computer.

Between Tap One and Tap Two, the mind has been demoted (No such thing) and explained (No such thing is a computer).

Whatever one might say about Steven Pinker's thoughts about the human mind, they do not lack for dramatic vigor.

ᴄᴂ

In 1936, the British logician Alan Turing published the first of his papers on computability. Using nothing more than ink, paper, and the resources of mathematical logic, Turing managed to create an imaginary machine, one capable of incarnating a very smooth, very suave imitation of the human mind.

Known now as a Turing machine, the device has at its disposal a tape divided into squares, and a reading head mounted over the tape. It has, as well, a finite number of physical symbols. The reading head may occupy one of a finite number of distinct physical states.

And thereafter the repertoire of its action is extremely limited. A Turing machine can, in the first place, recognize symbols, one square at a time. It can, in the second place, print symbols or erase them from the square it is scanning. And it can, in the third place, change its internal state, and move to the left or to the right of the square it is scanning, one square at a time.

There is no fourth place. Without a program a Turing machine can do nothing else. In fact, considered simply as a mechanism, a Turing machine can do nothing whatsoever, the thing existing in that peculiar world—my own, of course—in which everything is possible but nothing gets done.

Although imaginary at its inception, a Turing machine brilliantly anticipated its own realization in matter, with Turing's ideas giving rise to the modern digital computer.

The promotion of the computer from an imaginary to a material object serves the purpose of restoring it to the world that can be understood in terms of the physical sciences. As a physical device, nothing more than a collection of electronic circuits, the digital computer may be represented entirely by Clerk Maxwell's theory of the electromagnetic field. The distinction between a computer and its program is duplicated in the distinction between a physical system governed by certain specific laws and its initial condition—the state from which it starts. We are returned to the continuous and infinite world in which mathematical physics tracks the evolution of material objects moving through time in response to the eternal forces of nature itself.

<del>ᥴᠣ</del>

If something is gained by the assimilation of the brain to a computing device, something is lost as well, and that is the recognition that deep down every one of these metaphors is profoundly limited. A certain "power to alter things," Albertus Magnus remarked, "indwells in the human soul." The existence of this power is hardly in doubt. It is evident in every human act in which the mind imposes itself on nature by taking material objects from their accustomed place and rearranging them, and it is evident again whenever a human being interacts with a machine. Writing with characteristic concision in the *Principia*, Isaac Newton observed that "the power and use of machines consist only in this, that by diminishing the velocity we may augment the force, and the contrary." Although Newton's analysis was restricted to mechanical forces, his point is nonetheless general. A machine is a material object, a *thing*, and as such its capacity to do work is determined by the forces governing its nature *and* by its initial conditions. Before an inclined plane can do work, it must be *inclined*.

Those initial conditions must themselves be explained, and in the nature of things, they cannot be explained by the very device they serve to explain. An inclined plane does not incline *itself*. This is precisely the problem that Newton faced in the *Principia*. The magnificent system of the world that he devised explained why the orbit of the planets should be conic sections, but Newton was unable to account for the initial conditions that he had himself imposed on his system.

This pattern, along with its problem, recurs whenever machines are at issue, and it returns with a vengeance whenever computers are invoked as models for the human mind.

If the brain is a computer, then the very same thesis about the human mind should be in force whether we describe the human mind as a digital computer or whether we describe the human mind in terms of a device that is logically identical to a digital computer—an abacus, say. The thing is a trifle. Made of wood, it consists of a number of wires suspended in a frame and a finite number of beads strung along the wires. Nevertheless, an idealized abacus has precisely the power of a Turing machine, and so both the abacus and the Turing machine serve as models for a working digital computer. By parity of reasoning, they also both serve as models for the human mind.

Yet the thesis that the human mind is an abacus seems distinctly less plausible than the thesis that the human mind is a computer. And for an obvious reason: It is absurd. It is precisely when things have been reduced to their essentials that the interaction between a human being and a simple machine emerges clearly. That interaction is naked, a human agent handling an abacus with the same directness of touch that he might employ in handling a lever, a pulley, or an inclined plane.

With the nakedness of interaction revealed, a characteristic problem is revealed as well. While an abacus may represent certain human intellectual operations such as addition or subtraction, it cannot represent its own initial conditions. Regarding her big-nosed customers with indifference, it is the Chinese cashier at the Imperial Gardens who does that. The force that she brings to bear on an abacus is muscular, and so derived from the chemistry of the human body, these causes

ultimately emptying out into the great ocean of physical interactions whose energy loosens and binds the world's large
molecules.

No known chain of causes accommodates the inconvenient
fact that by setting the initial conditions of a simple machine,
a Chinese cashier has brought about a novel, an unexpected,
an entirely idiosyncratic distribution of matter. The initial
state of any mechanical artifact represents what the anthro-
pologist Mary Douglas called "matter out of place." Explaining even the simplest of human acts, the trivial tap or touch
that sets a polished wooden bead spinning down a wire, requires tracing the causal chain backward. But that leads only
to a wilderness of causes, each of them displacing material
objects from their proper settings, so that in the end the mystery is simply shoveled back until the point is reached when it
can be safely ignored.

A chain of physical causes is thus not obviously useful in
explaining how the human mind imposes itself on matter. But
neither does it help to invoke the hypothesis that another abacus is needed to fix the initial conditions of the first. If each
abacus requires yet another abacus in turn, the road lies open
to the madness of an infinite regress. Daniel Dennett has argued in *Brainstorms: Philosophical Essays on Mind and Psychology* that if receding computers are, like himself, diminished in
their capacity, the regress may end in some trivial mechanical
device, one that he describes as "stupid." But if those receding

computers are too thick to function as models of the mind, how do they do what they are said to do?

And if they are not, how have we been advanced?

If we are able to explain how the human mind works neither in terms of a series of physical causes nor in terms of a series of infinitely receding mechanical devices, what then is left? There is the ordinary, very rich, infinitely moving account of mental life that without hesitation we apply to ourselves. It is an account frankly magical in its nature. The human mind registers, reacts, and responds; it forms intentions, conceives problems, and then, as Aristotle dryly noted, it *acts*.

And in none of this do we seem to be doing anything that can be explained or expressed in terms of what the brain does, or what any machine can do.

"Mind is like no other property of physical systems," the physicist Erich Harth has reasonably remarked. "It is not just that we don't know the mechanisms that give rise to it. We have difficulty in seeing how any mechanism can give rise to it."

LAKE OF DOUBT

One of the curiosities about the current enthusiasm for various pseudo-scientific accounts of the human mind is that deep down those most willing to promote its premises are least willing to accept its conclusions.

Whatever scientists may say on those all too frequent occasions when they are advising the rest of us what to think, one thing that they do not say is that they believe what they are telling us to think. The result at times is moving. Writing to the widow of his old friend Michele Besso, Einstein remarked that "now he has departed from this strange world a little ahead of me. That means nothing. People like us, who believe in physics, know that the distinction between past, present, and future is only a stubbornly persistent illusion." Whatever the illusion, he acknowledged ruefully, it is one "stubbornly held."

More often than not, the disjunction between what scientific figures claim and what they believe represents a strikingly successful exercise in self-delusion. When it was first published, Richard Dawkins's *The Selfish Gene* took the intellectual world by storm. Conversion experiences among young men were widely reported. They still are. The idea that we are all "lumbering robots" designed by natural selection to advance the interests of our genes has become one of those things believed widely because widely believed. The mystery has even been celebrated in art. First promoted at the Cambridge Science Festival, *Lifetime: Songs of Life & Evolution* is a drama whose "mission [is] to spread the good word on evolution." There are tributes to Richard Dawkins, one song entitled "I'm a Selfish Gene and I'm Programmed to Survive." Although I have not seen it, I am persuaded that this theatrical endeavor is horrible beyond measure.

What is remarkable in all this is that no one taking selfish genes seriously takes them seriously. Richard Dawkins has gone out of his way to affirm that he, at least, is not under the control of his genes. "I too am an implacable opponent of genetic determinism," he has written. His genes are not so selfish as to tell him what to do. Who knows what might happen if he gave them a free hand? He may lumber, but if he does, the dead wood is under his control.

It is the rest of us who must lumber on.

The most unwelcome conclusion of evolutionary psychology is also the most obvious: If evolutionary psychology is true, some form of genetic determinism must be true as well. Genetic determinism is simply the thesis that the human mind is the expression of its human genes. No slippage is rationally possible.

Psychologists will now object. They have the floor. There is the environment, they say. It, too, plays a role. The environment has, of course, been the perpetual plaintiff of record in *Nurture v. Nature et al.* But for our purposes it may now be dismissed from further consideration. If the environment controls how men are made and how they act, then they are not born that way; and if they are not born that way, an explanation of the human mind cannot be expressed in evolutionary terms.

How could it be otherwise? On current views, it is the gene that is selected by evolution, and if we are not controlled by our genes, we are not controlled by evolution.

If we are not controlled by evolution, *evolutionary* psychology has no relevance to the origin or nature of the human mind.

And if it is has no relevance whatsoever to the origin and

nature of the human mind, why on earth is it promoted so assiduously to within an inch of its life or ours?

A successful evolutionary theory of the human mind would, after all, annihilate any claim we might make on behalf of human freedom. The physical sciences do not trifle with determinism: It is the heart and soul of their method. Were boron salts at liberty to discard their identity, the claims of inorganic chemistry would seem considerably less pertinent than they do.

When Steven Pinker writes that "nature does not dictate what we should accept or how we should live our lives," he is expressing a belief—one obviously true—entirely at odds with his professional commitments.

If ordinary men and women are, like Pinker himself, perfectly free to tell their genes "to go jump in the lake," why pay the slightest attention to evolutionary psychology?

Why pay the slightest attention to Pinker?

Either the theory in which he has placed his confidence is wrong, or we are not free to tell our genes to do much of anything.

If the theory is wrong, which theory is right?

If no theory is right, how can "the idea that human minds are the product of evolution" be "unassailable fact"?

If this idea is *not* unassailable fact, why must we put aside "the idea that man was created in the image of God"?

These hypotheticals must now be allowed to discharge themselves in a number of categorical statements:

There is *no* reason to pay attention to Steven Pinker.

We do *not* have a serious scientific theory explaining the powers and properties of the human mind.

The claim that the human mind is the product of evolution is *not* unassailable fact. It is barely coherent.

The idea that man was created in the image of God *remains* what it has always been: And that is the instinctive default position of the human race.

9

Miracles in Our Time

"I N MUCH the same way as prophets and seers and great theologians seem to have died out," Christopher Hitchens has claimed in *God Is Not Great*, "so the age of miracles seems to lie somewhere in our past."

Have they? Does it?

I would have thought that Einstein, Bohr, Gödel, Schrödinger, Heisenberg, Dirac, and even Richard Feynman were all in their own way prophets and seers.

Apparently not.

But miracles? The word seems to engender its own current of contempt. If one demands of a miracle that it violates the inviolable, there could be no miracles. It follows that there are none. Somehow this seems rather too easy a victory to afford even Christopher Hitchens a sense of satisfaction. No one is much concerned to debate the proposition that what could not be cannot be. Nor is it particularly invigorating to designate as a miracle an unexpected turn of events favoring oneself, as when a diagnosis proves benign or a divorce final. A miracle is what it seems: an event offering access to the divine. And if this is what miracles are, whether they are seen will, of course, always be contingent on who is looking. The miracles of religious tradition are historical. They reflect the power the ancient Hebrews brought to bear on their experiences. They did what they could. They saw what they could see. But we have other powers. We are the heirs to a magnificent scientific tradition. We can see farther than men whose horizons were bounded by the burning desert.

In a remark now famous, Richard Feynman observed with respect to quantum electrodynamics that its control over the natural world is so accurate that in measuring the distance from New York to Los Angeles, theory and experiment would diverge by less than the width of a human hair. Einstein's theory of general relativity is in some respects equally accurate. We cannot account for these unearthly results. The laws of nature neither explain themselves nor predict their success. We have no reason to expect such gifts, and if we *have* come to expect them, this is only because, as the saints have always warned, we expect far more than we deserve.

GOD OF THE GAPS

Scientific atheism is not an undertaking that has cherished rhetorical inventiveness. It has one brilliant insult to its credit, and that is the description of intelligent design as "creationism in a cheap tuxedo." I do not know who coined the phrase, but whoever it was, *chapeau*. By the same token, it has only one stock character in repertoire, and that is the God of the Gaps. Unlike the God of Old, who ruled irritably over *everything*, the God of the Gaps rules over gaps in argument or evidence. He is a presiding God, to be sure, but one with limited administrative functions. With gaps in view, He undertakes his very specialized activity of incarnating Himself as a stopgap. If He is resentful at the limitations in scope afforded by His narrow specialization, He is, scientific atheists assume, grateful to have any work at all.

When the gaps are all filled, He will join Wotan in Valhalla.

As a rhetorical contrivance, the God of the Gaps makes his effect contingent on a specific assumption: that whatever the gaps, they will in the course of scientific research be filled. It is an assumption both intellectually primitive and morally abhorrent—primitive because it reflects a phlegmatic absence of curiosity, and abhorrent because it assigns to our intellectual future a degree of authority alien to human experience. Western science has proceeded by filling gaps, but in filling them, it has created gaps all over again. The process is inexhaustible. Einstein created the special theory of relativity to accommodate certain anomalies in the interpretation of Clerk Maxwell's theory of the electromagnetic field. Special relativity led directly to general relativity. But general relativity is inconsistent with quantum mechanics, the largest visions of the physical world alien to one another. Understanding has improved, but within the physical sciences, anomalies have grown great, and what is more, anomalies have grown great *because* understanding has improved.

The God of the Gaps? I am prepared with the best of them to revile and denounce him. It is easy enough to do just that, one reason that so many scientists are doing it. But why not say with equal authority that for all we know, it is the God of Old who continues to preside over the bent world with His accustomed fearful majesty, and that He has chosen to reveal Himself by drawing the curtain on His own magnificence at precisely the place in which general relativity and quantum mechanics should have met but do not touch? Whether gaps in the manifold of our understanding reveal nothing more than the God of the Gaps or nothing less than the God of Old is hardly a matter open to rational debate.

It is in this respect discouraging time and again to see the matter discharged in a peevish display of vanity. In considering the possibility that the facts of biology might suggest an intelligent designer, *which surely they do,* Emile Zuckerkandl has found it difficult to contain his indignation. Writing in the journal *Gene,* he overflowed into epithets: "The intellectual virus named 'intelligent design.' . . . This virus certainly is a problem in the country. . . . the 'creationists' . . . have decided some years ago . . . to dress up in academic gear and to present themselves as scholars . . . laugh off this disguise. Their . . . erroneous beliefs are weighty reasons to keep them in check. . . . they try to foster on society . . . some enterprising superghost. Naïve members of the public . . . a comical invitation . . . the wrong foot—the only foot on which promoters of intelligent design can get around . . . peddled to the public. The minority of 'intelligent designers' who have any true interest in biology . . . The 'intelligent designers'' theme song . . . guided by a little angel . . . medieval in concept . . . an intellectually dangerous condition . . . the divine jumping disease. . . . humanity dug herself into 'faiths' like a blind leech into flesh and won't let go. . . . Feeding like leeches on irrational beliefs . . . offensive little swarms of insects . . . must be taken care of by spraying biological knowledge. . . ."

Darwinian biologists are very often persuaded that there is a conspiracy afoot to make them look foolish.

In this they are correct.

DULL, DUTIFUL, SO VERY DARWIN

There are times, I suspect, when even the most ardent among biologists suspects that enough is enough. The Old Boy is everywhere; he has long since ascended to the Pantheon; schoolchildren hymn his name, and while the man himself seems to have been sober, melancholy, and boring, his admirers have over the past twenty years or so succeeded in suggesting that his effulgence was such that had he been embedded in the ocean floor, sailors might for centuries unerringly navigate by his luster. If Richard Dawkins has not yet proposed renaming various English banknotes in Darwin's favor, this is only because of late he has been too busy counting them.

Enough *is* enough.

The effort by Darwinian biologists to promote Darwin is simply explained. Within the English-speaking world, Darwin's theory of evolution remains the only scientific theory to be widely championed by the scientific community and widely disbelieved by everyone else. No matter the effort made by biologists, the thing continues to elicit the same reaction it has always elicited: You've got to be kidding, right? There is wide appreciation of the fact that if biologists are wrong about Darwin, they are wrong about life, and if they are wrong about life, they are wrong about everything.

Mindful of what is at stake—everything—biologists resemble the war horses mentioned in Job 39: 19–25: "He saith among the trumpets, Ha, ha." If they cannot fight the battles at hand, they are eager to refight the battles they have won. For Eugenie Scott, Paul Gross, Barbara Forrest, Robert Pennock, or Lawrence Krauss, it is yet 1925. John Scopes is in the dock. Clarence Darrow is at his side. And in the small towns where the prairie winds blow, the forces of right thinking are still occupied in doing battle for men's souls.

Suspicions about Darwin's theory arise for two reasons. The first: the theory makes little sense. The second: it is supported by little evidence. In his very long posthumous treatise, *The Structure of Evolutionary Theory*, Stephen Jay Gould explained "the bare bones" of natural selection in this way: "Organisms enjoying differential reproductive success will, on average, be those variants that are fortuitously better adapted to changing local environments." Those variants that are "better adapted" are, of course, precisely those "enjoying differential reproductive success." What else could they be? Biologists believe that tautologies play an unsuspected role in scientific thought and are for this reason worthy of respect. Of course they do.

As one might expect, a theory whose assumptions are empty may be widely confirmed by evidence whose relevance is negligible. Just how and when do species arise? The standard view throughout much of the twentieth century has been that geographical barriers, such as a mountain range or an open body of water, are necessary to force an ancestral population to diverge.

In a study reported in the November 20, 2007, edition of *Science Daily*, Vicki Friesen, a professor of biology, observed: "While that model fits for many parts of the natural world, it doesn't explain why some species appear to have evolved separately, within the same location, where there are no geographic barriers to gene flow." And, indeed, some species *have* evolved separately within the same location. Friesen's own research indicated that the band-rumped storm petrel shares its nesting sites in sequence with other petrels. This result conflicts with the standard view. In *Origins*, Darwin himself argued for just this possibility.

I have every confidence in Dr. Friesen's research and no way in which to dispute it. I am not about to investigate the band-rumped storm petrel. It is her conclusion that must give pause. It is "exciting," she affirms, "to be able to verify Darwin's original theory!"

But no *theory* has been confirmed since every possibility has been justified. Speciation proceeds in the presence of geographic barriers, and it proceeds in their absence. The demand that the facts somehow support the theory may thus be treated as it so often is in Darwinian thought, and that is as an inconvenience.

If the facts are what they are, the past is what it is— profoundly enigmatic. The fossil record may be used to justify virtually any position, and often is. There are long eras in which nothing happens. The fire alarms of change then go off in the night. A detailed and continuous record of transition between *species* is missing, those neat sedimentary layers, as Gould noted time and again, never revealing precisely the phenomena that Darwin proposed to explain. It is hardly a matter on which paleontologists have been reticent. At the very beginning of his treatise *Vertebrate Paleontology and Evolution*, Robert

Carroll observes quite correctly that "most of the fossil record does *not* support a strictly gradualistic account" of evolution. A "strictly gradualistic" account is precisely what Darwin's theory demands: It is the heart and soul of the theory.

But by the same token, there are no laboratory demonstrations of speciation either, millions of fruit flies coming and going while never once suggesting that they were destined to appear as anything other than fruit flies. This is the conclusion suggested as well by more than six thousand years of *artificial* selection, the practice of barnyard and backyard alike. Nothing can induce a chicken to lay a square egg or to persuade a pig to develop wheels mounted on ball bearings. It would be a violation, as chickens and pigs are prompt to observe and often with indignation, of their essential nature. If species have an essential nature that beyond limits cannot change, then random variations and natural selection cannot change them. We must look elsewhere for an account that does justice to their nature or to the facts.

Although Darwin depicted natural selection as a force "daily and hourly scrutinizing" the biological world—a description that would equally designate the activities of the Holy Ghost—efforts to *measure* natural selection have been remarkably unforthcoming. In a research survey published in 2001, and widely ignored thereafter, the evolutionary biologist Joel Kingsolver reported that in sample sizes of more than one thousand individuals, there was virtually no correlation between specific biological traits and either reproductive success or survival. "Important issues about selection," he remarked with some understatement, "remain unresolved."

Of those important issues, I would mention prominently the question whether natural selection exists at all.

Computer simulations of Darwinian evolution fail when they are honest and succeed only when they are not. Thomas Ray has for years been conducting computer experiments in an artificial environment that he has designated Tierra. Within this world, a shifting population of computer organisms meet, mate, mutate, and reproduce.

Sandra Blakeslee, writing for the *New York Times,* reported the results under the headline "Computer 'Life Form' Mutates in an Evolution Experiment: Natural Selection Is Found at Work in a Digital World."

Natural selection found at work? I suppose so, for as Blakeslee observes with solemn incomprehension, "the creatures mutated but showed only modest increases in complexity." Which is to say, they showed nothing of interest at all. This *is* natural selection at work, but it is hardly work that has worked to intended effect.

What these computer experiments *do* reveal is a principle far more penetrating than any that Darwin ever offered:

There is a sucker born every minute.

෴

If Darwin's theory of evolution has little to contribute to the content of the sciences, it has much to offer their ideology. It serves as the creation myth of our time, assigning properties to nature previously assigned to God. It thus demands an especially ardent form of advocacy. In this regard, Daniel Dennett, like Mexican food, does not fail to come up long after he has gone down. "Contemporary biology," he writes, "has demonstrated *beyond all reasonable doubt* that natural selection—the process in which reproducing entities must compete for finite resources and thereby engage in a tournament of blind trial and error from which improvements *automatically* emerge—has the power to generate breathtakingly ingenious designs" (italics added).

These remarks are typical in their self-enchanted self-confidence. Nothing in the physical sciences, it goes without saying—*right?*—has been demonstrated beyond all reasonable doubt. The phrase belongs to a court of law. The thesis that improvements in life appear *automatically* represents nothing more than Dennett's conviction that living systems are like elevators: If their buttons are pushed, they go up. Or down, as the case may be. Although Darwin's theory is very often compared favorably to the great theories of mathematical physics on the grounds that evolution is as well established as gravity, very few physicists have been heard observing that gravity is as well established as evolution. They know better and they are not stupid.

I mention these obvious points not in order once again to abuse poor Dennett, an activity that I never weary of undertaking, but to make a point of my own. The greater part of the debate over Darwin's theory is not in service to the facts. Nor to the theory. The facts are what they have always been: They are unforthcoming. And the theory is what it always was: It is unpersuasive. Among evolutionary biologists, these matters are well known. In the privacy of the Susan B. Anthony faculty lounge, they often tell one another with relief that it is a very good thing the public has no idea what the research literature *really* suggests.

"Darwin?" a Nobel laureate in biology once remarked to me over his bifocals. "That's just the party line."

WHAT BIOLOGISTS TALK ABOUT WHEN THEY TALK ABOUT LIFE

In the summer of 2007, Eugene Koonin, of the National Center for Biotechnology Information at the National Institutes of Health, published a paper entitled "The Biological Big Bang Model for the Major Transitions in Evolution."

The paper is refreshing in its candor; it is alarming in its consequences. "Major transitions in biological evolution," Koonin writes, "show the same pattern of *sudden emergence of diverse forms at a new level of complexity*" (italics added).

Major transitions in biological evolution? These are precisely the transitions that Darwin's theory was intended to explain. If those "major transitions" represent a "sudden emergence of new forms," the obvious conclusion to draw is not that nature is perverse but that Darwin was wrong.

"The relationships between major groups within an emergent new class of biological entities," Koonin goes on to say, "are hard to decipher and do not seem to fit the tree pattern that, following Darwin's original proposal, remains the dominant description of biological evolution." The facts that fall outside the margins of Darwin's theory include "the origin of complex RNA molecules and protein folds; major groups of viruses; archaea and bacteria, and the principal lineages within each of these prokaryotic domains; eukaryotic supergroups; and animal phyla."

That is, pretty much everything.

Koonin is hardly finished. He has just started to warm up. "In each of these pivotal nexuses in life's history," he goes on to say, "the principal 'types' seem to appear rapidly and fully equipped with the signature features of the respective new level of biological organization. No intermediate 'grades' or intermediate forms between different types are detectable."

The phrase *intermediate forms* has a particular poignancy in context. It has been by an appeal to those intermediate forms that a very considerable ideology has been created. To doubt their existence is to stand self-accused. To go further and suggest that they are, in fact, *imaginary* evokes a frenzy of fearful contempt so considerable as to make civilized discourse impossible.

Koonin's views do not represent the views of the Darwinian establishment. If they did, there would be no Darwinian establishment. They are not uncontested. And it may well be that they are exaggerated. Koonin is nonetheless both a serious biologist and a man not well known for a disposition to self-immolation.

And in a much more significant sense, his views are simply part of a much more serious pattern of intellectual discontent with Darwinian doctrine. Writing in the 1960s and 1970s, the Japanese mathematical biologist Motoo Kimura argued that on the genetic level—the place where mutations take place— most changes are selectively neutral. They do nothing to help an organism survive; they may even be deleterious. A competent mathematician and a fastidious English prose stylist, Kimura was perfectly aware that he was advancing a powerful argument against Darwin's theory of natural selection. "The neutral theory asserts," he wrote in the introduction to his masterpiece, *The Neutral Theory of Molecular Evolution*, "that the great majority of evolutionary changes at the molecular level, as revealed by comparative studies of protein and DNA sequences, are caused *not by Darwinian selection* but by random drift of selectively neutral or nearly neutral mutations" (italics added).

This is radical doctrine. Waves of probability ebb and flow throughout the molecular structure of a living organism. Invisible to the scrutinizing force of natural selection, mutations drift through the currents of time. Whether a mutation is fixed within a population or whether it is simply washed away is a matter of chance.

The neutral theory of molecular evolution was never destined to achieve wide favor among Darwinian biologists.

Kimura's treatise is framed as a powerful but difficult mathematical argument. But population geneticists understood its importance, even if they disagreed in some of its details. To the extent that the neutral theory is true, Darwin's theory is not.

This has prompted at least certain population geneticists to deplore in print the sheer effrontery that is so conspicuous a feature of the popular literature devoted to Darwin's theory. Richard Dawkins has appeared as tempting a squab within the tent of population genetics as he has long seemed without. Writing in the *Proceedings of the National Academy of Sciences,* the evolutionary biologist Michael Lynch observed that "Dawkins's agenda has been to spread the word on the awesome power of natural selection." The view that results, Lynch remarks, is incomplete and therefore "profoundly misleading." Lest there be any question about Lynch's critique, he makes the point explicitly: "What is in question is whether natural selection is a necessary or sufficient force to explain the emergence of the genomic and cellular features central to the building of complex organisms."

But if it is quite possible that natural selection is neither necessary nor sufficient to account for the complexity of living systems, then it is also possible that it is of no relevance to living systems whatsoever.

The demotion of natural selection from biological superpower to ideological sad sack throws into bright relief an obvious question: How to explain on the basis of a random walk the startling coherence and complexity of living organisms? If the question is obvious, so, too, is its answer: We have no idea. "The general foundations for the evolution of 'higher' from 'lower' organisms," Emile Zuckerkandl has written, *"seems so far to have largely eluded analysis"* (italics added).

This is surely true. But the phrase *eluded analysis* conveys a current of intellectual optimism at odds with the facts. Something that has so far eluded analysis can hardly be assigned to a force that has so far eluded demonstration. It is in this context that Daniel Dennett's assertion that natural selection has been demonstrated "beyond all reasonable doubt" must be judged for what it is: It is the ecclesiastical bull of a most peculiar church, a cousin in kind to an ecclesiastical bluff. When Steven Pinker affirms that "natural selection is the *only* explanation we have of how complex life can evolve," he is very much in the inadvertent position of the apostles. Much against his will, he is bearing witness.

In all this, it is the reaction among the faithful that provokes no surprise. Within minutes of the publication of Koonin's paper, a call for censorship went up over the Internet. "Well," one solemn donkey wrote, "since it is clear that this paper will be on every ID/creationist blog on the planet in under 12 hours, I might as well put in my 2 cents early."

He might as well. And those two cents? What did they amount to?

One cent was devoted to a counsel of caution: "I think Koonin should give a little credit where credit is due to gradual, stepwise evolution."

The second cent was spent on a cry of alarm: "Sometimes you've got to wonder how many hangovers (i.e., creationist quote-mining and general confusion over the status of evolution outside of the specialist community, and needless wrangling within the specialist community) could be avoided if scientists would exercise just a little caution during the party (i.e., spending a little time soberly comparing their revolutionary ideas with more prosaic explanations)."

The words *if scientists would exercise just a little caution* have a meaning all their own. They are written in code. They convey the need, apparently imperative, for biologists to keep bad news to themselves.

What is left is the "general confusion" that the public so often suffers when it comes to Darwin and Darwinism. On this matter, biologists are not at all confused. Whatever the degree to which Darwin may have "misled science into a dead end," the biologist Shi V. Liu observed in commenting on Koonin's paper, "we may still appreciate the role of Darwin in helping scientists [win an] upper hand in fighting against the creationists."

It is hard to be less confused than that.

GREAT GAPS OF GOD

The God of the Gaps occupies a very considerable comfort zone in biology. He is right at home. We know better than we ever did that a great many aspects of biological behavior are innate. They arise in each organism. They are a part of its nature. This is certainly true of human beings. The point has been made with great force and plausibility by the linguist Noam Chomsky. Just as children are not taught to walk, they are not taught to speak. The environment serves only to trigger an innate maturational program. Human language is the very expression of human nature.

This is widely seen as offering dramatic confirmation of what Chomsky himself has called the "biological turn." It is surely easy to see why. What is innate in an organism, so it is claimed, reflects its genetic endowment, and its genetic endowment reflects the long process in which random variations were sifted by a stern and unforgiving environment. If we are born with the ability to acquire a natural language, the gift lies within our genes and our genes lie within the shifting tides of time.

This view is so common that it is often forgotten that it is also incoherent. What is both interesting and innate in an organism cannot be explained in terms of its genetic endowment. If the concept of a gene is given any content at all—not a certainty by any means—it is entirely with the context of molecular biology and biochemistry. The gene is a chemical, a part of the molecule deoxyribonucleic acid, or DNA. Its function is straightforward: It specifies the proteins needed by a living organism, and it species them by means of a remarkably complicated system of translation and transcription. To speak clearly of the genetic endowment of an organism is to speak *only* of the passage from one chemical structure to another— *that and nothing more.*

But to speak of the genetic endowment *of* an organism in terms that answer any interesting question *about* the organism is to go quite beyond the coordination of chemicals. It is to speak of what an organism does, how it reacts, what plans it makes, and how it executes them; it is to assign to a biological creature precisely the properties always assigned to such creatures: intention, desire, volition, need, passion, curiosity, despair, boredom, and rage.

These are not properties of a living system that can be easily seen as the consequences of any chemical reaction. It would be like suggesting that a tendency toward kleptomania follows the dissociation of water into hydrogen and oxygen. This may well be so. Research is required. But if it *is* so, it represents a connection that we do not understand and cannot grasp. The gap is too great. When Richard Dawkins observes that genes "*created* us, body and mind" (emphasis added), he is appealing essentially to a magical connection. There is nothing in any precise concept of the gene that allows a set of biochemicals to create anything at all. If no precise concept of the gene is at issue, the idea that we are created by our genes, body and mind, represents a far less plausible thesis than the correlative doctrine that we are created by our Maker, body and mind.

c6⤴

"The more comprehensible the universe becomes," Steven Weinberg has written, "the more it also seems pointless." I suspect that Professor Weinberg is not actively called upon when victims of life's injuries require solicitude. Beyond demanding that they deal with it, what could he say? This has struck many as an ungenerous attitude, and Weinberg has made every effort to cover his comment in confusion, chiefly by observing after the fact that he considers the universe a fine place, after all. If Weinberg's power, prestige, and intellectual authority are not evidence in favor of the universe, then at least he can say that he has gotten from it a very good deal.

My sympathies are nonetheless with the old, sour, unregenerate Weinberg. He had a point. The arena of the elementary particles—*his* arena—*is* rather a depressing place, and if it resembles anything at all it rather resembles a fluorescent-lit bowling alley seen from the interstate, tiny stick figures in striped bowling shirts jerking up and down in the monstrously hot and humid night.

What *is* its point?

We seem to live our lives in perfect indifference to the Standard Model of particle physics, the world we inhabit not only remote from the world it describes but different in detail, thank God.

Over *there*, fields are pregnant with latent energy, particles flicker into existence and disappear, things are entangled, and no one can quite tell what is possible and what is actual, what is here and what is there, what is now and what was then. Solid forms give way. Nothing is stable. Great impassive symmetries are in control, as vacant and unchanging as the eye of Vishnu. Where they come from, no one knows. Time and space contract into some sort of agitated quantum foam. Nothing is continuous. Nothing stays the same for long, except the electrons, and they are identical, like porcelain Chinese soldiers. A pointless frenzy prevails throughout.

Over *here*, space and time are stable and continuous. Matter is what it is, and energy is what it does. There are solid and enduring shapes and forms. There are no controlling symmetries. The sun is largely the same sun now that it was four thousand years ago when it baked the Egyptian deserts. Changes appear slowly, but even when rapid, they appear in stable patterns. There is dazzling variety throughout. The great river of time flows forward. We anticipate the future, but we remember the past. We begin knowing we will end.

The God of the Gaps may now be invited to comment—strictly as an outside observer, of course. He is addressing us. And this is what He has to say: You have *no* idea whatsoever how the ordered physical, moral, mental, aesthetic, and social world in which you live could have ever arisen from the seething anarchy of the elementary particles.

It is like imagining sea foam resolving itself into the Parthenon.

And even though He is speaking strictly as an observer, perhaps He will be forgiven for asking of Christopher Hitchens, who has wandered into this discussion prepared to dispute anyone at the bar, "Where wast thou when I laid the foundations of the earth? declare, if thou hast understanding."

᠅

These examples may be multiplied at will. They form a common pattern, one in which a mystery is in evidence, but one demanding for its resolution intellectual insights that we do not possess and cannot honestly say we will in time command. No one has the faintest idea whether the immense gap between what is living and what is not may be crossed by any conceivable means. It is therefore no surprise that the National Academy of Sciences has taken pains to affirm that it has already been crossed. "For those who are studying aspects of the origin of life, the question no longer seems to be whether life could have originated by chemical processes involving non-biological components but, rather, what pathway might have been followed." The view among biochemists actively engaged in research is different. "The de novo appearance of oligonucleotides on the primitive earth," Gerald F. Joyce and Leslie Orgel remarked in their chapter of a volume entitled *The RNA World*, "would have been a near miracle." Oligonucleotides are among the indispensable building blocks of living systems.

A *near* miracle is a term of art. It is like a near miss. And a miss, it should be recalled, is as good as a mile.

The theories that we have do what they can do, *and then they stop*. They do not stop because a detail is missing; they stop because *we* cannot go on. Difficulties are accommodated by the magician's age-old tactic of misdirection.

Writing about the eye in *On the Origin of Species,* Darwin confessed that its emergence troubled him greatly. He was nonetheless able to resolve his own doubts in his favor, and ever since, biologists have assumed that inasmuch as Darwin proposed a solution, they need not face a problem. The solution that Darwin proposed and defended was simply to point to countlessly many examples of intermediate visual structures scattered throughout the animal kingdom. It formed an interesting argument. It did not touch the central issue. The eye is not simply a biological organ, although surely it is that. It is a biological organ that allows living creatures to *see.* If we cannot say what seeing comes to in physical or material terms, then we cannot say whether *any* theory is adequate to explain the appearance of an organ making sight possible.

This is precisely what we cannot say. The physical details are in part understood. Light strikes the eye in the form of photons but it exits the eye in terms of electrical signals. In between, bipolar cells convey visual information to ganglion cells, which in turn conduct information to the optic nerve. Thereafter the optic nerve conveys electrical signals to the brain. The brain in turn twitches into life, neurons firing here and there, the gooey mass for a moment convulsed.

And directly thereafter, *I* see the looming mass of Notre Dame, all gray stone and leering gargoyles, a long line of plodding tourists shuffling toward the door leading to the cathedral's towers, the horses of the National Guard dropping their straw-filled waste in the center of the street as they clip-clop patiently toward their stables, the light, the hot haze, dust dancing in the air.

I open my eyes and my eyes are filled.

How do the twitching nerves, chemical exchanges, electrical flashes, and computational routines of the human eye and brain provide a human being with his *experiences?*

The gap opened between causal sequences that with a moving finger we can trace from one point to the next and the light-enraptured awareness to which they give rise is unfathomably large because it spans an incommensurable distance. The processes involved in sight are biological, chemical, and in the end physical. It may well be that at some point in the future, a physicist, using quantum electrodynamics perhaps, might be in a position to write down their equations. Whether such an equation will encompass our experiences—why, this is something we simply do not know.

"Today we cannot see whether Schrödinger's equation contains frogs, musical composers, or morality," Richard Feynman remarked in his lectures on turbulence. The remark has been widely quoted. It is honest.

The words that follow are rarely quoted. "We cannot say whether something beyond it like God is needed, or not. And so we can all hold strong opinions either way."

These words form an obvious inferential chain. If we do not know whether Schrödinger's equation will one day accommodate our experience, we certainly do not know whether our experiences reflect anything less than a miracle.

For the moment, if asked to stand and declare ourselves on the most elementary aspects of the world in which we live—*We see it*—we can say nothing.

TIME, DEATH, LIFE, AND LONGING

For almost as long as the physical sciences have made their claims, poets and philosophers have observed that there is something inhuman about the undertaking they represent. They are right. We gain purchase on the physical world first by stripping it to its simplest form, and second by emptying it of its emotional content. Whatever the elementary particles may be doing, they are not forming political alliances, or looking on one another with mute incoherent longing, or casting an anxious eye on the clock, or waking with a start in the early hours of the morning, wondering what it all means, or coming to realize that they are destined to fall like the leaves of the trees leaving not a trace behind.

These are things *we* do: It is in our nature to do them. But how do we do them? By what means accessible to the imagination does a sterile and utterly insensate physical world become the garrulous, never-ending, infinitely varied, boisterous *human* world? The more the physical world is studied, and the richer our grasp of its principles, the greater the gap between what it represents and what we embody.

In 1948, Kurt Gödel provided a subtle argument for the thesis that time does not exist. In the course of providing a new solution of Einstein's equations for general relativity, Gödel showed that the universe might be rotating in a void, turning serenely like a gigantic pinwheel. In a universe of this sort, each observer sees things as if he were at the center of the spinning, with the galaxies—indeed, the whole universe—rotating about him. As the galaxies rotate, they drag space and time with them, like propeller blades pulling water in their wake. A rotating universe turns space and time around in spirals. By moving in a large enough circle around an axis, at something approaching the speed of light, an observer might catch his own temporal tail, returning to his starting point at some time earlier than his departure.

If time moves in circles, and an observer can return to his own past, it seems to follow that effects might be their own causes.

Gödel recognized that rotating universes may be physically unrealistic, but they are possible, and once seen as possibilities, they cannot be unseen. Within these universes, time is an illusion. If time is an illusion in some universe, then features of time that we take for granted in our universe must be either accidents or gifts.

If time is an accident, it is inexplicable, and if a gift, it is unexpected. These conclusions, as Gödel remarked dryly, "can hardly be considered satisfactory."

When, in 1948, Gödel first published his thoughts, the

reaction was polite, but indifferent. Einstein appreciated his friend's genius but thought his theories bizarre. But to read the literature of theoretical physics almost sixty years later is to be struck by the extent to which, at the far reaches of speculation, very similar ideas are reappearing, almost as if they were caught in one of those strange vortices that, in Gödel's view, returned things to the past. Edward Witten and Alain Connes have both speculated that in the end, space and time might not have been there in the beginning. They are not necessary features of the physical world. When the deepest theories of physics are finally set out, perhaps centuries from now, they will not mention space and time. God knows if they will mention anything that we can understand.

We live by love and longing, death and the devastation that time imposes. How did they enter into the world? And why? The world of the physical sciences is not our world, and if *our* world has things that cannot be explained in *their* terms, then we must search elsewhere for their explanation.

We may allow ourselves in the early twenty-first century to neglect the Red Sea and to regard with unconcern the various loaves and fishes mentioned in the New Testament. We who are heirs to the scientific tradition have been given the priceless gift of a vastly enhanced sense of the miraculous. This is something that the very greatest scientists—Newton, Einstein, Bohr, Gödel—have always known and always stressed.

We are where human beings have always been, conveyed by miracles and yet unsure of the conveyance, unable to place our confidence completely in anything, or our doubt completely in everything.

When asked what he was in awe of, Christopher Hitchens responded that his definition of an educated person is that you have some idea how ignorant you are. This seems very much as if Hitchens were in awe of his own ignorance, in which case he has surely found an object worthy of his veneration.

CHAPTER

10

The Cardinal and His Cathedral

I N DECEMBER 1613, a full sixty years after the death of
Nicolaus Copernicus, the earth still stood at the center of
the universe. It had not moved, and it had not been moved.
Occupying distinguished positions in all the great universities
of Europe, sophisticated astronomers saw no reason to dilute
their faith in the ancient Ptolemaic system. It had stood the test
of time, and it was accurate. The view that the earth was in motion around the sun they rejected because it seemed an offense
to intuition and common sense. And so it was. To the obvious
question why the earth's motion was not readily discernible,
Copernican astronomy could offer no credible response.

Five years later, the Church placed Copernicus's treatise,

De revolutionibus orbium coelestium (On the revolutions of the celestial spheres), on the index of banned books. In 1633, the Roman Inquisition placed Galileo Galilei on trial. He stood trapped, clever sniping Jesuits badgering him to renounce his view that the earth but not the sun was in motion. His tormenters capered and danced. In the end, Galileo *did* renounce his heretical doctrines, but he remained inwardly defiant. *Eppur si muove*, he was heard to mutter to himself when the proceedings concluded.

Yet it moves.

At least, this is the story that has been handed down to us. It is a tale that has engendered a long-standing myth of clerical ignorance and religious intolerance.

The facts are rather different, as the facts so often are.

⤚❦

Intoxicated by the new astronomical theories advanced by Copernicus and Johannes Kepler, and often helping himself to their ideas without bothering overmuch to credit their influence, Galileo had in 1613 committed his thoughts about science, religion, and astronomy to paper in a letter to his friend the Benedictine Benedetto Castelli. His letter is a great soulful cry, a plea for tolerance and freedom of inquiry. It is as well one of the governing documents of the modern scientific era, a kind of legal charter.

Galileo begins by assenting to a proposition that he proposes almost at once to deny: "The Holy Scripture can never lie or err, and . . . its declarations are absolutely and inviolably true." This is on its face an odd claim, even if in the context of early-seventeenth-century intellectual life it was a matter of orthodoxy, for it seems to conflate three quite different ideas. The first, that certain texts can never *lie*; the second, that they can never *err*; and the third, that they are not only true but *absolutely* true. But texts—written words, after all—can *neither* lie *nor* err, although they can certainly convey a lie or communicate an error. Lying and erring are things that men and women do. Texts can, on the other hand, be true or false, but Galileo is concerned to repeat the common view that biblical texts are not only true, but true absolutely and inviolably. And this suggests that such texts express propositions that not only are true, *but could not be false.*

Now, Galileo's scientific career was, if nothing else, a matter of demonstrating that in certain fundamental respects, the ancient and subtle Ptolemaic system, according to which the heavens revolved around the earth in a series of celestial spheres, was mistaken. But the Ptolemaic account *was* the biblical account. It was, in fact, the account common in the ancient Near East, where only the Greeks were daring enough to speculate that the earth might be in motion around the sun, and even the Greeks were unable to reconcile this thesis with the plain evidence of their senses. They were not, after all, flying into space from its surface, and if the earth was in motion, why weren't they? Thus, as Galileo perfectly well understood, biblical inerrancy and the claims advanced by Copernicus and

Kepler stood in conflict. An irresistible force had encountered an immovable object.

The friction thus engendered, Galileo proposed to ameliorate by means of a semantic dodge. "Although the scriptures cannot err," he wrote, "nevertheless some of its interpreters and expositors can err in various ways." Such errors typically involve the confusion of metaphorical and literal meaning. Taken literally, scriptures would seem to assign to God "feet, hands and eyes," and this, Galileo assumes, is quite absurd, although he makes this assumption by means of no argument. Moslem theologians of the tenth century had, after all, argued the contrary with great heat and no little eloquence.

There then follows a passage of quite extraordinary importance, one that has worked its way through every part of our own scientific and secular culture: "Thus given that the Scripture is not only capable but necessarily in need of interpretation different from the apparent meaning of [its] words, it seems to me that in disputes about natural phenomena, it should be reserved to the last place." This opinion, although provocative in the context of seventeenth-century thought, is today uncontroversial. The sentences that follow are otherwise: "For the Holy Scripture and Nature both equally derive from the divine Word, the former as the dictation of the Holy Spirit, the latter as the most obedient executrix of God's commands." Although inspired by the Holy Spirit, scriptures belong to the world of *appearances,* and appearances can be confused or misleading. With nature, things are completely different. "Nature is inexorable and immutable," Galileo writes, "and she does not care at all whether her recondite reasons and modes of operation are revealed to human understanding, and so she never transgresses the terms of the laws imposed on her." What Galileo calls "sensory experiences placed before our eyes or necessary demonstrations concerning nature" have an intrinsic force denied scripture itself, and in a conflict between the two, it is nature that must prevail.

This *is* revolutionary doctrine, and in Galileo's mind, one revolution engenders another. "Philosophy is written in this grand book of the universe," he affirms, his words again canonical, "which stands continually open to our gaze. But the book cannot be understood unless one first learns to comprehend the language and read the alphabet in which it is composed."

From this remarkable declaration, it follows that Nature *is* a book; and from what Galileo has already written, it follows that "nature never transgresses the terms of the laws imposed upon her."

These assertions imply that the Book of Nature is inerrant, so that the doctrine of biblical inerrancy, a staple of Christian thought, has not at all been discarded in Galileo's mind, but *transferred*. A new, greater, grander book now occupies his attention, but even though new, greater, and grander, the Book of Nature—*the* Book—is nonetheless very much like the old book. It is *inerrant*.

The "book of God's word" and the "book of God's works," Francis Bacon argued, are not in conflict.

How could they be?

They are the *same* book.

᧚

Hearing that unorthodox opinions were afoot, a Dominican, Niccolò Lorini, expressed his scruples in a letter written on February 7, 1615, to Cardinal Paolo, prefect of the Holy Office in Rome. Galileo's letter, he declared, was "suspicious or presumptuous." In order "to show their cleverness," Galileo and his followers were "airing and scattering broadcast [i.e., making known] in our steadfastly Catholic city, a thousand saucy and irreverent surmises." Lorini had earlier admitted to Galileo that he knew nothing of mathematics or physics, and in words that even today compel admiration, admitted that he knew even less about this "Ipernic or whatever his name is." He was, of course, referring to Copernicus.

And then a Carmelite named Paul Anthony Foscarini thought to compose a letter of his own, entitled "Copernicus and the Motions of the Earth and the Immobility of the Sun." It was, in fact, less a letter and more of a tract, a vigorous defense of the new astronomy. If mathematical physics and Holy Scripture were in conflict on certain matters, Foscarini suggested, then so much the worse for Holy Scriptures. And they *were* plainly in conflict.

Exultavit ut gigas currendam viam, the Psalmist had written about the sun.

"He rejoiceth like a giant to run the way."

Foscarini had persuaded himself that his enthusiasm was infectious without ever once worrying that it might be contagious. He sent a copy of his letter to Robert Cardinal Bellarmine.

᭡

An engraving of Cardinal Bellarmine by the Flemish artist Valdor of Liege depicts a man of about fifty. The cardinal is wearing a red hat, the sign of his office, and his shoulders are sheathed in red clerical robes. His face suggests a man one would be glad to know but unwilling to cross—careful eyes, an aquiline nose, and round, rubicund cheeks descending smoothly into a smooth, trimmed Vandyke beard. The forehead is creased and the edges of his eyes are crinkled, but not in any way indicating amusement. The man is plainly a prince of the Church, familiar with power and accustomed to human vanity. When Church officials commented on his outstanding piety and almost supernatural goodness—he was said to be fond of the poor—they did so in order to justify denying him the papacy. He is today a saint, circumstances suggesting that at his trial, the Devil's Advocate was indisposed.

Receiving Foscarini's letter in 1615, Bellarmine sent a response that arrived on April 12.

"My very dear Reverend Father," Bellarmine begins suavely, and afterward I paraphrase. It has been a pleasure for me to read your letter. It exhibits such skill and learning.

Bellarmine's praise was not insincere. He had, various stories indicated, once looked through a telescope pointed inconclusively toward Saturn, and he had seen enough so that with ringed traces of the eyepiece raccooning his eyes, he had muttered something indicating his pleased astonishment.

Nevertheless, the tone of Bellarmine's letter now changes. He will be brief, he informs this provincial rustic. No doubt Foscarini has little time to read, but more to the point, *he* has little time to write.

The Copernican assumption, the cardinal affirms, that it is the sun that stands still and the earth that moves might well "save appearances," and so conform to the facts better than the ancient Ptolemaic theory, with its wearisome eccentrics and epicycles. Let us say that this is so. "There is in this," the cardinal allows, "nothing dangerous."

But to go further into the frank affirmation that the sun really *is* immovable and the earth really *is* in motion—this, Bellarmine declares, "*is* a very dangerous thing."

Sixteen years before, Bellarmine had served the Church as an inquisitor at the trial of Giordano Bruno, one of history's lamentable pests, and Bruno had been burned at the stake, Cardinal Bellarmine approving the verdict and having done nothing to prevent its execution. When written by a man prepared to put other men to death, the words *very dangerous* have a force they might not otherwise possess. The cardinal, one imagines, has caught Foscarini's attention.

"Whenever a true demonstration would be produced that the sun stands at the center of the world"—and none has been vouchsafed me, the cardinal is quick to affirm—"then at that time it would be necessary to proceed with great caution in interpreting Scriptures which seem to be contrary."

This is so very reasonable as to place in doubt the very idea of clerical intolerance. Bellarmine is arguing, after all, only that in matters of astronomy, judgment might be suspended and *not* that inquiry must be stopped.

But suppose, the cardinal continues, the conflict between astronomical fact and Holy Scripture should prove irremediable; suppose, in fact, that a demonstration—not a conjecture, not an assumption, not one of these, please forgive me, Your Reverence, *amusing* suppositions that are so prominently a feature of your letter—were made available that the sun is *in fact* immovable.

Yes, suppose just that.

The cardinal now contemplates this appalling possibility with all his intellectual sophistication. If it came to that, he writes—if the sun really does stand still—"it would be better to say that we do not understand Holy Scripture than to say that what has been demonstrated is false."

But this is, of course, precisely what Galileo had urged— a grand, quite self-conscious project of avoiding conflict by feigning confusion.

⚘

The passionate drama played out four hundred years ago is playing out again. And why not? The characters that it involved are a part of the human comedy. If in the seventeenth century, the cardinal was willing to say that we might have misunderstood religion in order to uphold science, in the twenty-first, he is willing to say that we might have misunderstood science in order to uphold religion. It is Western science that is *our* church, the place in which we repose our confidence and our trust. I am among the faithful. And I am devoted to the church. I have, after all, spent my life studying its texts.

Far more than Isaac Newton—implacable, remote, incomprehensible in his genius—Galileo Galilei has entered contemporary life as the very soul and symbol of a way of thought. He is intensely human, and for this reason, sympathetic. He gave way before the Roman Inquisition but in the end he got his way. The Western world now thinks in his terms. We have for more than three hundred years occupied a Galilean universe.

Wir müssen wissen, wir werden wissen, the great German mathematician David Hilbert affirmed in an address given in 1930.

We must know, we will know.

The long Galilean moment in the history of thought is coming to an end. Shortly after Hilbert delivered his address, Kurt Gödel demonstrated that mathematics was inherently incomplete. If science in the twentieth century has demonstrated anything, it is that there are limits to what we can know. What we might wish and what we can have are not necessarily the same. A far older view of human life has entered a position of authority in our affairs.

At the very same moment that Hilbert grandly affirmed his

program of intellectual conquest, Galileo's other heirs were completing the last of the revolutions in physical thought. The Standard Model of particle physics is their monument. And thereafter there has been nothing. There has been nothing, that is, that could properly be expressed in Galilean terms.

Niccolò Lorini, so eager to denounce what he could not understand or did not wish to grasp, is also a familiar figure: He is destined now and forever to sound twittering notes of alarm with respect to doctrines that he finds alarming.

It hardly matters which doctrines have provoked his alarm; poor Niccolò is prepared to denounce them all. If in the seventeenth century they were scientific but not religious, in the twenty-first century they are religious but not scientific. Niccolò may today be found wherever the faith is under attack. Darwin's theory of evolution is the obvious example, because Darwin's theory is virtually the only part of church teaching commonly understood. It may be grasped by anyone in an afternoon, and often is. A week suffices to make a man a specialist. The great virtue of Darwin's theory, Richard Dawkins has argued, is that it has made it possible to be an intellectually fulfilled atheist. Dawkins's claim, while it has been widely repeated, has not been widely believed. "Two-thirds of Americans," the *New York Times* reported, "say that creationism should be taught alongside evolution in public schools." But even among those quite persuaded of Darwin's theory, "18 percent said that evolution was 'guided by a supreme being.'"

Under these circumstances, freedom of thought very often

appears as an inconvenience to those, like Niccolò Lorini, with a position to protect and enemies on all sides. A paper published recently in the *Proceedings of the Biological Society of Washington DC* concluded that the so-called Cambrian explosion, the sudden appearance of new life forms about 530 million years ago, could best be understood in terms of an intelligent design—hardly a position unknown in Western thought. The paper was, of course, peer-reviewed by three prominent evolutionary biologists. Wise men attend to the publication of every one of the *Proceedings'* papers, but in the case of Stephen Meyer's "The Origin of Biological Information and the Higher Taxonomic Categories," the Board of Editors was at once given to understand that they had done a bad thing. Their indecent capitulation followed at once.

Publication of the paper, they confessed, was a mistake. It would never happen again. It had barely happened at all.

"If scientists do not oppose antievolutionism," remarked Eugenie Scott, the executive director of the National Center for Science Education, "it will reach more people with the mistaken idea that evolution is scientifically weak." Scott's understanding of "opposition" had nothing to do with reasoned discussion. It had nothing to do with reason at all. *Discussing* the issue was out of the question. Her advice to her colleagues was considerably more to the point: "Avoid debates."

There is nothing surprising in any of this. I myself believe that the world would be suitably improved if those with whom I disagree were to lapse into silence.

↶

There is finally Cardinal Bellarmine; he is today where he was in the seventeenth century, and that is within the shadows, a man disposed to display his hand only when his hand is forced. If he is on those occasions useful in virtue of the sliver of pure steel running through his character, his usefulness is circumscribed by his quivering intelligence. Stern as a defender of the faith, he is, in his heart of hearts, a witness to its limitations.

The cardinal speaks today to those whose faith is sincere but whose doubts are significant. He speaks for *me,* and I suppose that in seeing something sympathetic in the cardinal, I have in return spoken for him.

No less than other men, the cardinal understands that in the twenty-first century, the symbol and the glory of faith is the cathedral that science has constructed from its great physical theories. The thing is immense. It can be seen from every vantage point, and even those ill at ease in its presence cannot escape its shadow.

But the cathedral is now some four hundred years old. The walls have aged into ocher and umber. Within, statues of all the saints stand on their pedestals. Newton is there, noble, uncorrupted, and aloof; and so, too, Clerk Maxwell, Albert Einstein, Niels Bohr, Werner Heisenberg, Erwin Schrödinger, Max Born, and Paul Dirac; Richard Feynman is the last. There are no others. There are *no* young saints and none have been proposed.

With a peasant smile of satisfaction creasing his narrow Italian face, the cardinal very much enjoys the grand spectacle that every day takes place within the cathedral and on the plaza on which it has been built. There are architects carrying rolled-up designs under their arms, masons stirring wet loads of cement, bricklayers, and carpenters; and hanging like monkeys from their scaffolds, stonecutters carving out gargoyles on all the high ledges.

But the cathedral is not finished. The interiors are crudely appointed. While some windows glow in subtle colors, others have been put in place before they have been stained, and in some parts of the great vault, simple pine boards have been nailed onto window frames still lacking any windows at all.

Although workmen speaking any number of languages may be seen every day on the cathedral's work site, there is a certain disorganization to their affairs. It is hardly surprising, given the fact that almost every worker belongs to a separate guild. Guild officials have been known to bring work to a halt over the most trivial of circumstances.

When the cathedral was first proposed many long years ago, the great visionaries imagined a single unified and compelling structure, its massive walls embracing a serene volume of space and light, its flanks ascending smoothly upward so that a slender spire emerged naturally to pierce the sky. Sketches of the original cathedral may still be found in the cathedral's basement, where mice have taken over all the filing cabinets.

The spire has not been built, and in the clear moonlight,

the cathedral looks unbalanced, almost as if it were a cripple defiantly waving a stump against the sky. The rumor is current among knowledgeable architects that from the first, the cathedral was constructed from incompatible blueprints.

The towers do not quite match. One is austere and classical. The other ornamented and baroque.

How was this overlooked?

At the very top of the cathedral, where the spire is intended to pierce the sky, but where only a small stub now exists, workmen have put down their tools. They do not know how to proceed. The architects are of little help. They consult their drawings, but the more their drawings occupy their attention, the less they are able to determine what they mean.

The cardinal longs to see the spire finished, thrust into the sky gleaming, so that *he* can step back and see it soar.

But the spire presents any number of difficult problems. Some of them are financial. Like every cathedral, this one is supported by public funds. Very often, the cardinal finds himself pleading for money before various church groups. It is a role he finds distasteful. Who would not?

There is dissension among the architects. Some now argue for a spire that is taller than the one planned; others for one that is shorter. And some believe that it should remain an idea, one that all men can see, without ever being translated into stone.

Catching himself in these visions of grandeur, the cardinal reminds himself that cathedrals have been known to collapse, and in thinking of the weight of the faith he has invested in the cathedral, he wonders—it is only natural—whether any structure can support such weight.

Although a visionary, the cardinal is also a practical man. He believes in costs and is apprehensive about expenses. A design should really be tested by experiment. The architects have said so. But the spire is projected to weigh tons and cost millions.

How *could* it be tested?

And if it could be tested, by what means could the test be tested?

What a question, the cardinal reflects. How can faith be tested? What *is* its test?

To discontinue work on the cathedral is unthinkable, the cardinal reflects, but even he does not know whether the spire will ever be built. No one is sure. It is possible that the cathedral will *forever* remain incomplete.

Every now and then careless tourists with no sense of its weight in history dismiss the cathedral as so much antique stone. What is its point? They snap pictures, and they are gone.

How little they understand.

But *does* the cathedral have a point?

Standing before the cathedral to which he has devoted his life, the cardinal says at least this to himself: that it has given meaning to those who have worked on it, and satisfaction to those who worship in its dim interior.

No one could bear its loss. It has become a monument, and

when from the plaza the professional beggars and sly tradesmen and rouge-lipped prostitutes look up, they see that great looming *familiar* thing, as natural as the space that contains it and the space that it contains.

From time to time, the cardinal allows himself to be questioned by the faithful. He is courteous, polite, and reserved. But he is distant.

"Your Eminence," they ask in every language of the world, "does our cathedral support the faith by which it is supported?"

The cardinal smiles enigmatically, a sly, ironic, distant, tender smile. Standing there on the cathedral's steps, he pauses to reflect, the light glinting from his miter, and his hooded eyes troubled.

He does not answer, but if he did, this is what he would say:

Does *any* cathedral?

ACKNOWLEDGMENTS

I am grateful to Ann Coulter for having brought the idea for this book to the attention of Crown Forum.

And I am grateful to my editor, Jed Donahue, and my agent, Susan Ginsburg, for the careful reading they gave the manuscript.

Many of the ideas that I present in this book were first expressed in essays that I have written for *Commentary* over the past ten years. I am deeply indebted to Neal Kozodoy for affording me the hospitality of his journal, and for his own incisive and very often skeptical reaction to what I wrote.

It is a special pleasure to record my indebtedness to the Discovery Institute for loyal support over many years. That the institute has been vilified by all the right people is a special sort of satisfaction.

INDEX

abacuses, 173, 174
Abu Tammam, 5
agnosticism, 2
Albertus Magnus, 173
Al-Ghazâli, Abu Hamid
    Muhammad, 16, 17
Anthropic Principle, 126–36,
    152
anti-Semitism, 28–31
apes, 155–56, 158, 160–65
Arabs, 13–17, 28
Aristotle, 65, 175
astronomy, 13–14, 209–17
atheism:
    argument for, 137–53
    as ideology, 6
    literary, 4–5, 11–12
    militant, 2, 5
    scientific, 3–4, 18, 20–21, 27, 33,
        34, 36, 43, 133, 183
    Thomas Aquinas and, 66–67
    village, 2–3
Atkins, Peter, 3, 7, 95–96
atoms, 53, 73

Auden, W. H., 109
Avalos, Hector, 18, 52

Bacon, Sir Francis, 2, 213
Bava Mezia, 29
Bell, John, 94
Bellarmine, Robert Cardinal,
    215–17, 221
Bethell, Tom, 8
"Beyond Belief" conference
    (2007), 20–21
Bible, 71, 84, 103, 130, 146, 207,
    213
Big Bang, 69–74, 78–81, 86,
    97–98, 101–4, 114
"Biological Big Bang Model for
    the Major Transitions in
    Evolution, The" (Koonin),
    192–93, 196, 197
"biological turn," 198
biology, 31–32, 184, 197–99
    see also evolutionary biology
Blackburn, Simon, 35, 38, 133

ABOUT THE AUTHOR

DAVID BERLINSKI has a Ph.D. from Princeton University and has taught mathematics and philosophy at universities in the United States and France. He is the bestselling author of such books as *A Tour of the Calculus*, *The Advent of the Algorithm*, and *Newton's Gift*. A senior fellow at the Discovery Institute in Seattle and a former fellow at the Institute for Applied Systems Analysis and the Institut des Hautes Études Scientifiques, Berlinski writes frequently for *Commentary*, among other journals. He lives in Paris.